United States Nuclear Regulatory Commission

Protecting People and the Environment

Fuel Behavior under Abnormal Conditions

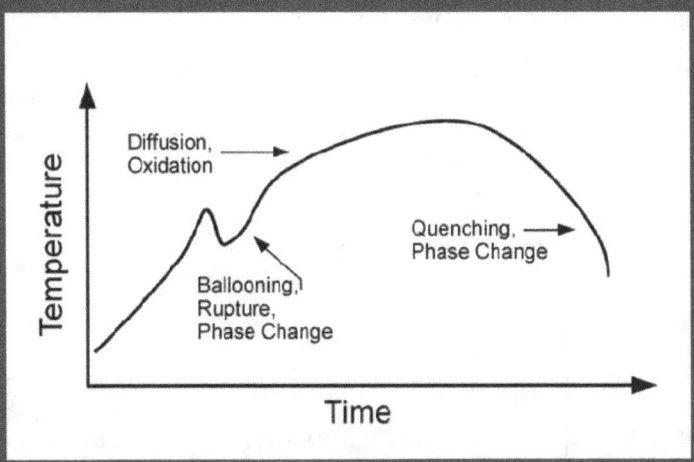

NUREG/KM-0004
January 2013
Office of Nuclear Regulatory Research

AVAILABILITY OF REFERENCE MATERIALS
IN NRC PUBLICATIONS

NUREG/KM-0004

United States Nuclear Regulatory Commission

Protecting People and the Environment

Fuel Behavior under Abnormal Conditions

Manuscript Completed: April 2009
Date Published:

Prepared by
R.O. Meyer

Prepared for
Office of Nuclear Regulatory Research
U.S. Nuclear Regulatory Commission
Washington, DC 20555-0001

ABSTRACT

Under normal operating conditions, cladding and core structural materials operate around 300 degrees Celsius (C), and fuel pellets experience peak temperatures below 2,000 degrees C at the pellet centerline. At these temperatures, fuel cladding integrity is maintained, and fission products are contained within the fuel rods. However, under some abnormal conditions, higher temperatures and other conditions significantly alter the behavior of these materials. These conditions can threaten core coolability and lead to fission product release. This NUREG report considers the following two types of accident conditions: (1) reactivity-initiated accidents and (2) loss-of-coolant accidents. This report describes the fuel behavior of each accident condition from basic concepts to the current state of the art. It also mentions safety criteria, references the classic experimental work in each of these areas, and presents equations and figures that permit some quantitative evaluations.

CONTENTS

ABSTRACT iii

CONTENTS v

FIGURES vii

FOREWORD ix

ABBREVIATIONS xi

1. INTRODUCTION 1

2. REACTIVITY-INITIATED ACCIDENTS 3

   2.1   Reactivity-Initiated Accidents with Fresh Fuel 3

   2.2   Reactivity-Initiated Accidents with High-Burnup Fuel 9

3. LOSS-OF-COOLANT ACCIDENTS 15

   3.1   Loss-of-Coolant Accidents with Fresh Fuel 16

       3.1.1   Phase Change 16

       3.1.2   Ballooning and Rupture 16

       3.1.3   Oxidation 19

       3.1.4   Diffusion of Oxygen into the Metal 21

       3.1.5   Hydrogen Absorption 25

   3.2   Loss-of-Coolant Accidents with High-Burnup Fuel 26

       3.2.1   Effect of Hydrogen 26

       3.2.2   Inside Diameter Oxygen Pickup 27

       3.2.3   Fuel Relocation 31

4. REFERENCES 35

FIGURES

Figure 1 Dependence of pulse width on energy (fuel enthalpy change) for beginning-of-cycle and end-of-cycle conditions in PWRs (Figure 4 in Ref. 2) 4

Figure 2 Enthalpy of UO_2 as a function of temperature to 4,000 K (Figure 2-3 in Ref. 3) 5

Figure 3 Linear thermal expansion of UO_2 as a function of temperature 5

Figure 4 Significant rise in the cladding temperature after the power pulse (calculated with FRAPTRAN, Ref. 4) 6

Figure 5 High-temperature damage in the cladding of fresh fuel after its exposure to a large RIA power pulse (285 cal/g) in the Power Burst Facility test reactor (Figure 73 in Ref. 5 and Figure 5 in Ref. 6) 7

Figure 6 Photographs of fresh fuel rods after RIA testing in the Special Power Excursion Reactor Test (SPERT) reactor (Figure 1 in Ref. 7) 8

Figure 7 Maximum fuel enthalpy change for an RIA in a PWR at hot zero power for various values of delayed-neutron fraction, beta (Figure 15 in Ref. 2) 8

Figure 8 (a) Scanning electron microscope image showing the ultrahigh-burnup rim structure and (b) elemental distribution as a function of the distance from the fuel-to-cladding interface in a fuel rod with a burnup of about 105 GWd/t (Figure 1 in Ref. 8) 10

Figure 9 Edge-peaked pellet temperatures early in an RIA transient and center-peaked temperatures after significant heat transfer (based on Figures 24 and 25 in Ref. 9) ...11

Figure 10 Thermal expansion algorithms for (a) normal operation (highest temperature at pellet centerline) and (b) a zero-power RIA (highest temperature at the edge) in high-burnup fuel 12

Figure 11 Plastic strain measured in nonfailed cladding as a function of maximum fuel enthalpy change for tests in the CABRI test reactor and the NSRR (Figure 7 in Ref. 2) 12

Figure 12 Solubility of H in Zr as a function of temperature (based on Ref. 10) 13

Figure 13 RIA test data for irradiated fuel, plotted as maximum fuel enthalpy change as a function of H in the cladding (H content was estimated when data were not available); solid symbols indicate cladding failure (based on Figure 3 of Ref. 2) 14

Figure 14 Cladding temperature (solid line) and rod pressure (dashed line) during a postulated LOCA (Figure 6 in Ref. 11) 15

Figure 15 Illustration of temperature progression and associated phenomena during a postulated LOCA 16

Figure 16 Correlations of rupture temperature and corresponding cladding stress for three heating rates (degrees Celsius per second) from data for Zircaloy-4 heated in steam (Figure 3 in Ref. 12) 17

Figure 17 Maximum circumferential strain as a function of rupture temperature for Zircaloy cladding heated in steam (Figure 8 of Ref. 12) 18

Figure 18 Ballooned and ruptured region of unirradiated Zircaloy-2 after undergoing LOCA conditions (Figure 217 in Ref. 13) 18

Figure 19 Linear power generation from the metal-water reaction for cladding with a diameter of 1.25 centimeters and various oxide thicknesses (Figure 4-69 in Ref. 3) 21

Figure 20 Pseudobinary Zircaloy-oxygen phase diagram (Figure 4 in Ref. 16)........................22

Figure 21 Microscopic image of unirradiated Zircaloy-2 after oxidation in steam at 1,200 degrees Celsius for 600 seconds at room temperature (Figure 4 in Ref. 13)..23

Figure 22 Qualitative diagram of oxygen concentration in Zircaloy cladding exposed to steam at high temperature 23

Figure 23 Diagram of ring-compression test...24

Figure 24 Illustration of diffusion mechanisms in Zr-based cladding alloys 25

Figure 25 Older E110 cladding showing early breakaway oxidation at 1,000 degrees C compared with modern M5 cladding, which is resistant to breakaway oxidation (based on Figure 88 in Ref. 13 and on related unpublished figures) 26

Figure 26 Embrittlement threshold for cladding specimens exposed to steam at 1,200 degrees Celsius and quenched after slow cooling to 800 degrees Celsius (Ref. 19) 27

Figure 27 Interaction bonding layer between UO_2 fuel and Zircaloy-4 cladding in a high-burnup specimen at 83 GWd/t (Figure 10 in Ref. 20) 28

Figure 28 Free energy of formation of oxides of Zr-alloy constituents and some impurities (Figure 1 in Ref. 13)...29

Figure 29 Oxygen sources for diffusion into cladding metal during a LOCA 30

Figure 30 Grain morphology and alpha layer at the OD and ID of a high-burnup fuel rod exposed to LOCA conditions (Supplementary Figure 3.5.24 in Ref. 20)...................31

Figure 31 Fuel pellet fragments in PWR fuel rods after normal operation to low and medium burnup levels (Figure 30 in Ref. 21)..32

Figure 32 Cross-section of the rupture region of an irradiated fuel rod after testing under LOCA conditions (Figure 13 in Ref. 21) 32

Figure 33 Pellet stack reduction as a function of the increase in volume in cladding balloons for preirradiated fuel rods (Figure 32 in Ref. 21) 33

Figure 34 Gamma scan of very-high-burnup (approximately 92 GWd/t) fuel rod showing major loss of fuel material after LOCA testing (Figure 15 in Ref. 22)........................34

FOREWORD

As part of the U.S. Nuclear Regulatory Commission's (NRC's) knowledge management effort, one of our retiring staff members, Dr. R.O. Meyer, prepared textbook material regarding the subjects on which he worked during his entire 35-year career with the NRC. The following NUREG report presents this material, which discusses fuel rod behavior during postulated accidents that the NRC considers in safety analyses. Fuel behavior under conditions of reactivity-initiated accidents and loss-of-coolant accidents are timely subjects because the NRC is in the process of revising Regulatory Guide 1.77, "Assumptions Used for Evaluating a Control Rod Ejection Accident for Pressurized Water Reactors," issued May 1974, and Title 10 of the *Code of Federal Regulations* (10 CFR) 50.46, "Acceptance Criteria for Emergency Core Cooling Systems for Light-Water Nuclear Power Reactors." These subjects are discussed comprehensively and bring the reader current with the state of the art.

During his career, Dr. Meyer worked extensively in these areas and published more than 40 related papers and NUREG reports. His effort in documenting some of his knowledge in such a readable fashion is greatly appreciated.

ABBREVIATIONS

B-J	Baker-Just
BIGR	fast pulse graphite reactor
BWR	boiling-water reactor
C	Celsius
C-P	Cathcart-Pawel
cal/g	calories per gram
CFR	*Code of Federal Regulations*
ECR	equivalent cladding reacted
FRAPTRAN	Fuel Rod Analysis Program—Transient
GWd/t	gigawatt day per ton
H	hydrogen
H_2O	water
He	helium
ID	inside diameter
IGR	pulse graphite reactor
J/kg	joules per kilogram
K	Kelvin
LOCA	loss-of-coolant accident
LWR	light-water reactor
Nb	niobium
NRC	U.S. Nuclear Regulatory Commission
NSRR	Nuclear Safety Research Reactor
OD	outer diameter
PCMI	pellet-cladding mechanical interaction
psi	pounds per square inch
Pu	plutonium
PWR	pressurized-water reactor
RIA	reactivity insertion accident
Sn	tin
SPERT	Special Power Excursion Reactor Test
UO_2	uranium dioxide
wt.ppm	weight parts per million
Zr	zirconium
ZrO_2	zirconium dioxide

1. INTRODUCTION

Under normal operating conditions, cladding and core structural materials operate around 300 degrees Celsius (C), and fuel pellets experience peak temperatures below 2,000 degrees C at the pellet centerline. At those temperatures, zirconium (Zr) alloys remain in their low-temperature alpha phase (hexagonal close-packed structure). Furthermore, because the vapor pressures of major fuel constituents remain low, most fission products are retained within the fuel pellets. However, under some abnormal conditions, higher temperatures and other conditions significantly alter the behavior of the materials.

Several of these abnormal conditions are of special interest in addressing the safety of nuclear power reactors. Licensing authorities generally require that accidents be postulated, that safety equipment be installed to mitigate these accidents, and that analyses be performed to demonstrate that the plant would survive these accidents without major releases of radioactivity. As a result, a great deal of research has been conducted over the years in the United States as well as other nations possessing nuclear power technology, by both regulators and industry stakeholders, to study the behavior of fuel materials under abnormal conditions.

From the point of view of fuel behavior, reactivity initiated accidents (RIAs) and loss-of-coolant accidents (LOCAs) are the two most challenging types of accidents postulated for light-water reactors (LWRs). The RIAs of particular interest are the most severe overpower events, and the LOCAs of interest are the most severe undercooling events that seem credible. As such, these accidents involve the most limiting conditions that fuel materials may experience.

2. REACTIVITY-INITIATED ACCIDENTS

A large reactivity excursion initiates the postulated RIA. It is followed, in turn, by a power excursion, the expulsion of hot fuel particles into the coolant, and the conversion of that thermal energy into mechanical energy resulting in a steam explosion. The RIA is the type of accident that occurred at the Chernobyl plant in April 1986; the subsequent steam explosion in this accident ejected core materials and started a fire. An RIA would initiate differently in a boiling-water reactor (BWR) than it would in a pressurized-water reactor (PWR), but the progression of the subsequent events would be similar.

In a BWR, the accident is postulated to occur after a control blade hangs up in the core as the drive assembly is withdrawn, and some time later the control blade drops out of the core at its maximum velocity. This type of accident is called a rod-drop accident. An accident of this type that occurs under cold zero-power conditions produces the largest reactivity excursion. Although no rod-drop accidents have occurred in BWRs, the broken and cracked shafts found in the Oskarshamn and Forsmark plants in 2008 may be considered precursors to this type of accident (Ref. 1).

In a PWR, the accident is postulated to occur after a crack develops around an upper head penetration (nozzle) that houses a control rod drive, and the drive unit and its attached control rod cluster are ejected under pressure when the crack fails. This type of accident is called a rod-ejection accident. An accident of this type that occurs under hot zero-power conditions produces the largest reactivity excursion. This type of accident is not possible in a PWR under cold conditions because the vessel must be pressurized (and therefore hot) to cause the ejection. No rod-ejection accidents have occurred in PWRs either, but a precursor occurred in 2002 at the Davis-Besse Nuclear Power Station when a control rod drive nozzle remained attached only by its stainless steel liner following nozzle cracking and severe corrosion of the reactor head.

2.1 Reactivity-Initiated Accidents with Fresh Fuel

Regardless of the speed of a falling BWR control blade or an ejecting PWR control rod, the nuclear reaction soon becomes prompt critical. From that point on, the rate of power increase and subsequent Doppler-caused decrease depend only on nuclear properties of the core design. In general, the resulting pulse has a half-width (i.e., fuel width at half maximum) of only a few tenths of a millisecond. Because little heat is transferred from the fuel pellet to the cladding during this time, the event can be considered adiabatic. Consequently, a power pulse in an RIA is usually characterized by the maximum fuel enthalpy in calories per gram (cal/g) and by the pulse half-width in milliseconds. Figure 1 shows the pulse width for typical RIA enthalpy additions in a PWR (Ref. 2). For a given energy, the pulse is somewhat broader in a BWR because the nuclear coupling in that reactor type is looser than that of a PWR.

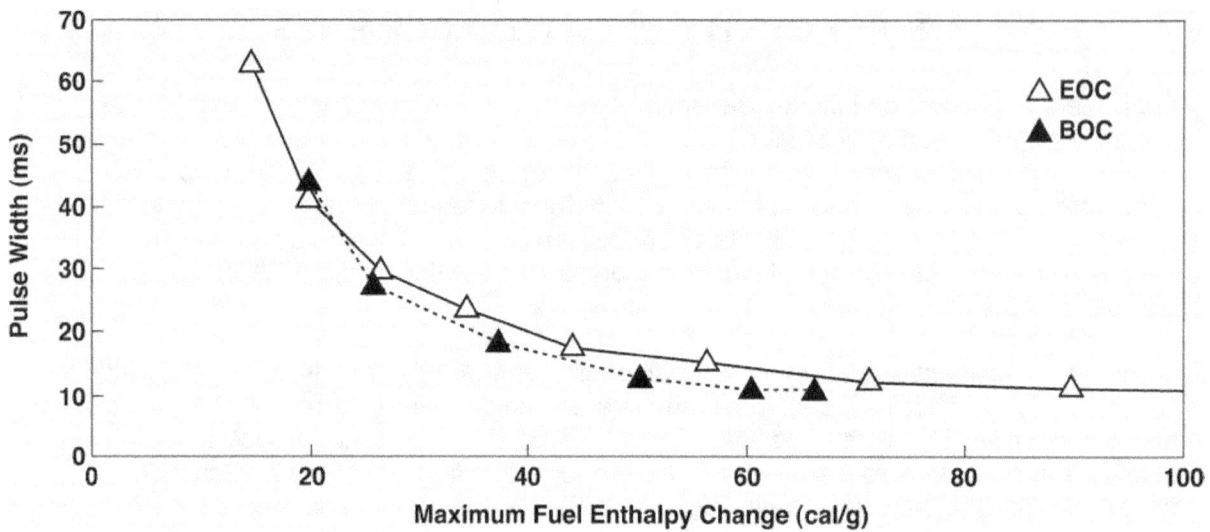

Figure 1 Dependence of pulse width on energy (fuel enthalpy change) for beginning-of-cycle and end-of-cycle conditions in PWRs (Figure 4 in Ref. 2)

Consider a pulse that deposits 200 cal/g (0.84×10^6 joules per kilogram (J/kg)) in the uranium dioxide (UO_2) fuel pellets within a Zircaloy-clad fuel rod, which is at zero power and room temperature. Assuming that the energy is deposited uniformly in the UO_2 pellets, the temperature of the pellets will rise to 2,600 Kelvin (K) (2,327 degrees Celsius) according to the enthalpy relation shown in Figure 2 (Ref. 3). From the thermal expansion relation shown in Figure 3 (Ref. 3), the linear thermal expansion of UO_2 going from room temperature to 2,600 K (2,327 degrees Celsius) is 2.9 percent. For a 15x15 fuel design, the as-fabricated gap between the pellets and the cladding is about 2 percent of the cladding diameter. The gap thus accommodates much of the pellet expansion, leaving a net strain in the cladding of less than 1 percent, which fresh Zr-alloy cladding can readily accommodate. Therefore, the mechanical interaction between the pellets and the cladding will not lead to cladding failure (i.e., through-wall penetration or fracture) for fresh fuel; however, the result is very different for high-burnup fuel (see Section 2.2).

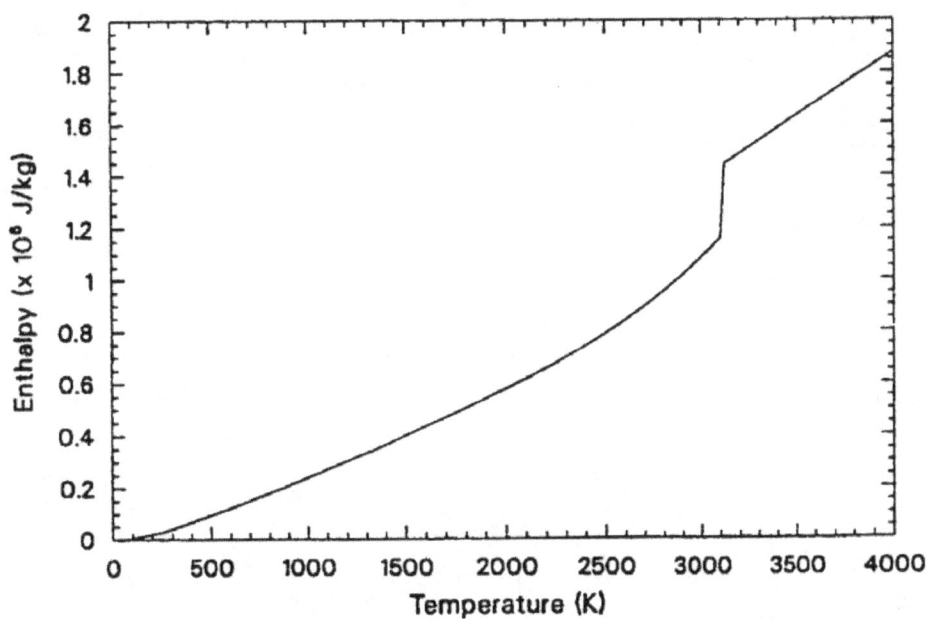

Figure 2 Enthalpy of UO2 as a function of temperature to 4,000 K (Figure 2-3 in Ref. 3)

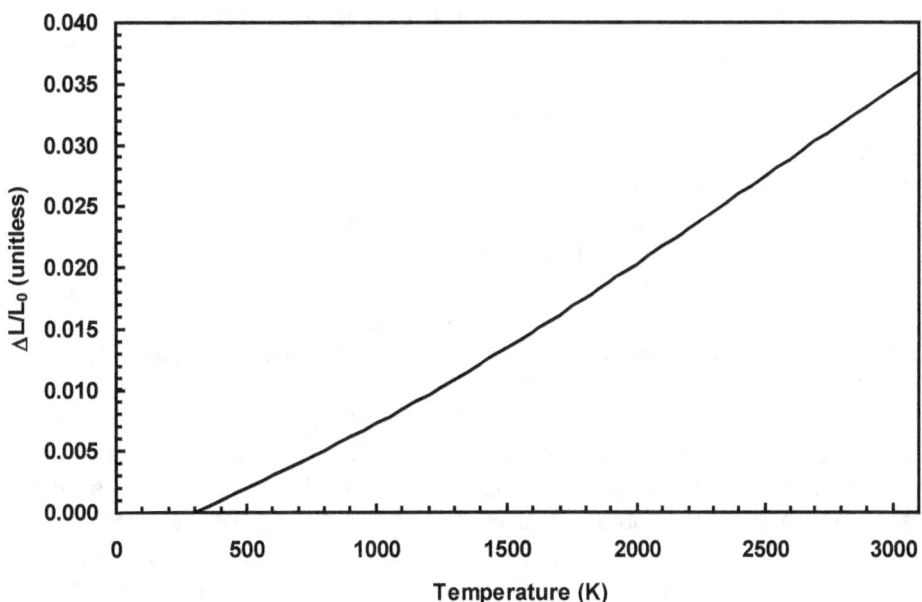

Figure 3 Linear thermal expansion of UO2 as a function of temperature (Equation 2-36 in Ref. 3)

However, note that 2,327 degrees C in this example is well above the melting point of Zr (1,852 degrees C). Although the cladding temperature cannot rise significantly during the 10-millisecond period of the power pulse, the heat in the pellets eventually goes through the cladding into the coolant. During this process, the cladding temperature rises significantly as shown in Figure 4, which was calculated with FRAPTRAN (Ref. 4) for a somewhat smaller pulse (108-cal/g peak fuel enthalpy) and includes appropriate heat-transfer kinetics.

NUREG/KM-0004

Figure 4 Significant rise in the cladding temperature after the power pulse (calculated with FRAPTRAN, Ref. 4)

Actual cladding temperatures depend on heat transfer across the pellet-to-cladding gap and from the outside of the cladding to the coolant; cladding temperatures also depend on time, thermal conductivities, and heat capacities. Because these processes are complicated and difficult to model, the progression of fuel damage during RIAs is customarily determined empirically by testing. Figure 5 shows the result of one such test on a fresh fuel rod. This test was performed by the Department of Energy for the Nuclear Regulatory Commission in the Power Burst Facility test reactor at an initial temperature of 265 degrees Celsius and resulted in a peak fuel enthalpy of 285 cal/g with a pulse width of 13 milliseconds (Ref. 5).

Based on the photomicrograph in Figure 5, very high cladding temperatures were clearly achieved. The beta phase seen in the center of the cladding is evidence of temperatures above the phase transformation at about 850 degrees C. Also, the large oxide layers on the inside and outside surfaces of the cladding could not have developed during the transient unless temperatures were well above 1,000 degrees C. Furthermore, the eventual failure (quench fracture) is a brittle fracture as deduced from the fracture surface, and oxygen diffusing into the bulk metal from the oxide on the surfaces caused cladding embrittlement. This amount of embrittlement in such a short time indicates temperatures well above 1,200 degrees C. The high-temperature behavior of this Zircaloy cladding is thus similar to that which occurs in a LOCA (see Section 3). Finally, the failure clearly occurred at a low temperature during the cooldown phase because there is no oxide present on the fracture surface.

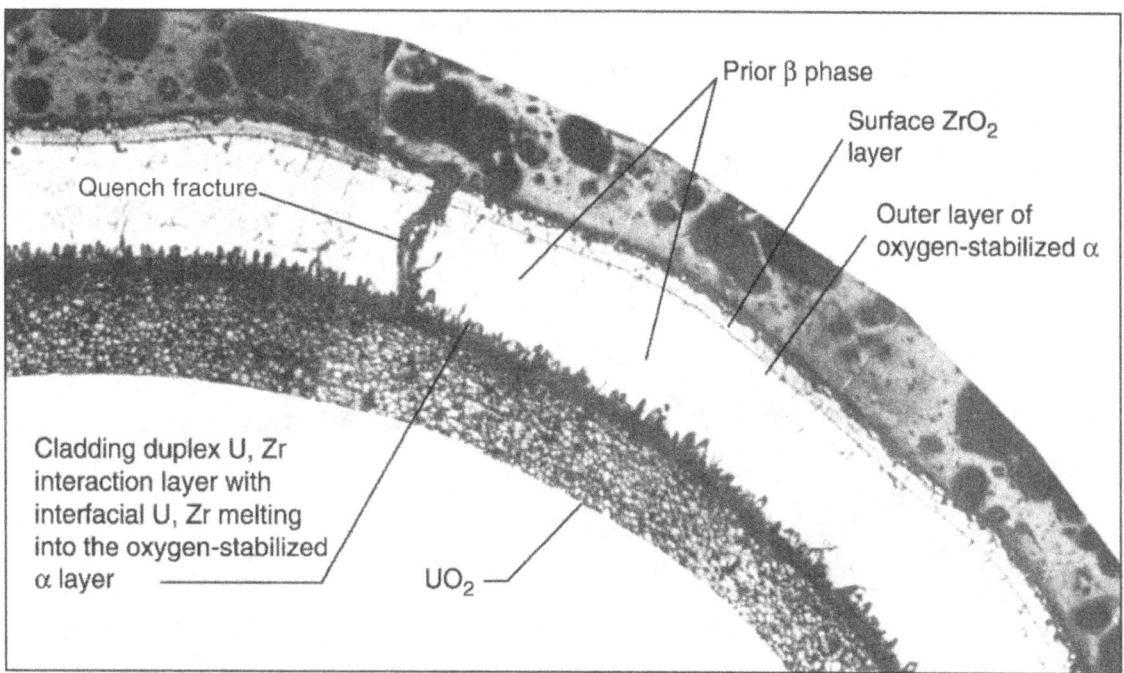

Figure 5 **High-temperature damage in the cladding of fresh fuel after its exposure to a large RIA power pulse (285 cal/g) in the Power Burst Facility test reactor (Figure 73 in Ref. 5 and Figure 5 in Ref. 6)**

The nuclear industry has used a traditional value of 170-cal/g total enthalpy to indicate the level at which cladding failure will occur as a result of overheating. Although the technical basis for this traditional value is apparently lost, the value is consistent with the test results from the pulsed graphite reactor (IGR) and the fast pulse graphite reactor (BIGR) in Russia; these results indicate that swelling and rupture, rather than a mechanical mechanism, caused cladding failure (Ref. 2).

Although cladding failure is of some interest in assessing the safety consequences of an RIA because of the small releases of fission products, the major consequence of this type of accident comes from a possible steam explosion. Merely causing a crack in the cladding will not necessarily expel hot fuel particles into the coolant such that a steam explosion could occur. However, Figure 2 shows that fuel enthalpies greater than 275 cal/g (1.15×10^6 J/kg) would result in some molten UO_2. At some enthalpy higher than this value, the thermal expansion that takes place during the phase change would cause the expulsion of molten or hot solid particles of UO_2 from the rod into the coolant. MacDonald et al. (Ref. 7) found damage levels for fresh and low-burnup fuel in their classic work, as shown in Figure 6, and concluded that a safety limit should have been set at about 230 cal/g rather than the 280-cal/g limit that the U.S. Nuclear Regulatory Commission (NRC) adopted because of confusion between total energy deposition and peak fuel enthalpy (i.e., some energy is lost in the tests even in the first 10 milliseconds).

Total Energy Deposition (cal/g)		Peak Fuel Enthalpy (cal/g)

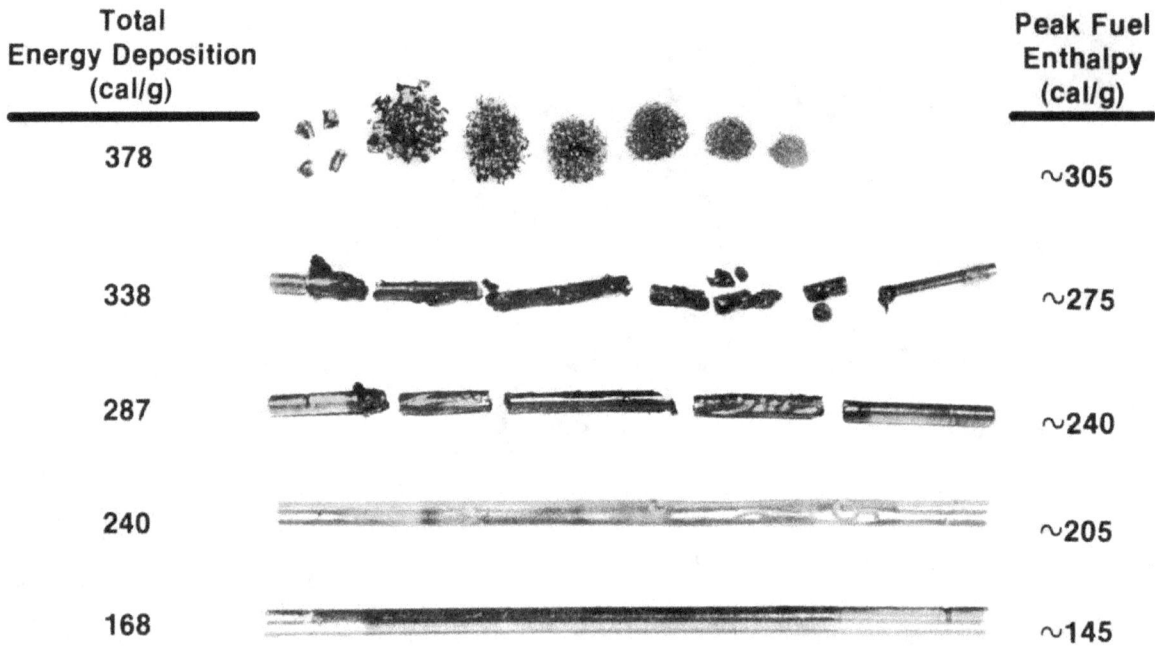

378		~305
338		~275
287		~240
240		~205
168		~145

Figure 6 Photographs of fresh fuel rods after RIA testing in the Special Power Excursion Reactor Test (SPERT) reactor (Figure 1 in Ref. 7)

Fuel enthalpy resulting from an RIA is related to the worth of the ejected control rod, as shown in Figure 7 (Ref. 2). Because control rod worth is generally less than $2, Figure 7 shows that the fuel enthalpy change would usually be less than about 70 cal/g. If 25 cal/g is added for an RIA that starts at about 300 degrees Celsius in a PWR, the total is still less than the 170-cal/g enthalpy level at which cladding failure would occur and much less than the 230-cal/g level at which hot fuel particles might be expelled from a fuel rod. Therefore, no cladding failure, fuel dispersal, or steam explosion would be expected for fresh LWR fuel in an RIA.

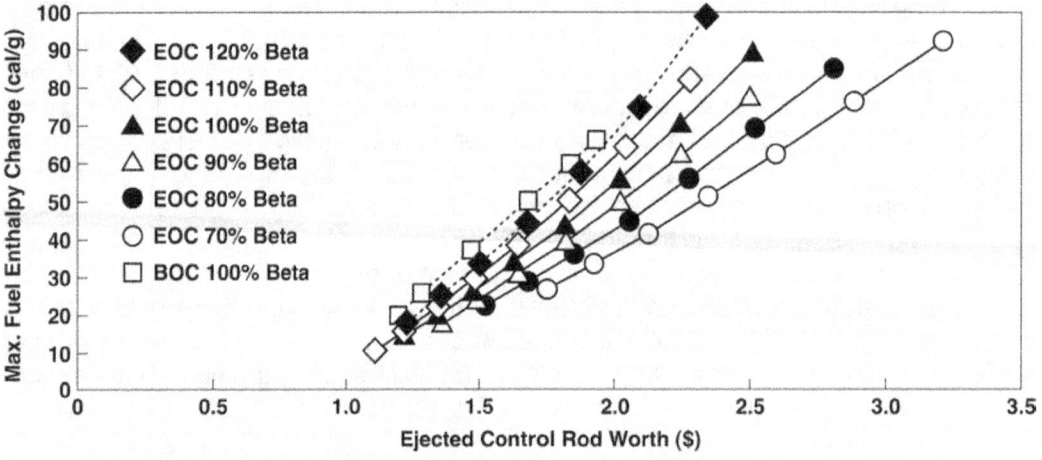

Figure 7 Maximum fuel enthalpy change for an RIA in a PWR at hot zero power for various values of delayed-neutron fraction, beta (Figure 15 in Ref. 2)

2.2 Reactivity-Initiated Accidents with High-Burnup Fuel

At the present time, power reactors are achieving fuel burnups as high as a little over 60 gigawatt day per ton (GWd/t) (rod average), and burnups above 40 GWd/t are considered high. Several properties of high-burnup fuel radically alter the evolution of an RIA in both BWRs and PWRs. In particular, the fissile atom distribution is changed within the fuel pellets, the pellet-to-cladding gap is closed, hydrogen (H) is present in the cladding, and fission gas bubbles are found within the fuel pellets, as can be seen in Figure 8(a).

Figure 8(b) shows fissile atom redistribution (Ref. 8) in a fuel rod with a very high burnup. Natural shielding suppresses fissions in the center of the fuel pellets relative to that in the peripheral regions. This effect becomes greatly enhanced as burnup progresses by the buildup of plutonium (Pu), which is largely fissile, near the pellet surface. The concentration of Pu atoms in this particular high-burnup fuel rod more than doubles near the pellet surface relative to that at the centerline, thus resulting in very high fuel temperatures near the pellet surface early in the power transient (Figure 9). This near-surface peak temperature is much higher than the peak temperature in a fresh fuel rod.

High fuel surface temperatures substantially enhance the thermal expansion of the pellet. Figure 10(a) illustrates a typical thermal-expansion algorithm that is used in most computer models, especially for normal operating conditions when the peak temperature is at the pellet centerline. In this algorithm, the centerline node expands more than the other nodes do, and the expansion of each node is added to the next. Because UO_2 has no significant tensile strength, each nodal ring cracks, and the incremental expansions adequately add together without much sensitivity to the number of rings chosen.

Figure 10(b) shows the thermal-expansion algorithm that must be used for an RIA with high-burnup fuel. This figure shows that the outer node expands more than the other nodes do, and the expansion decreases from the surface to the center. Conceptually, the outer node determines the expansion of the pellet as if the whole pellet were at the peak temperature. This algorithm poses two practical problems. First, the temperature that is assigned to the outer node will be sensitive to the thickness of that node and hence the number of nodes chosen. Second, a very thin outer node will not have the ability to deform the cladding outward without some inward deformation of the node itself. Although this makes analysis difficult, a high-burnup fuel pellet will apparently expand considerably more than a fresh or low-burnup pellet for a given fuel enthalpy in an RIA.

The next factor to consider in high-burnup fuel is the pellet-to-cladding gap. This gap closes as a consequence of fuel pellet swelling during normal operation combined with cladding creepdown caused by a pressure differential during normal operation. As a result, all expansion of the fuel pellet will be transmitted to the cladding without any accommodation as in fresh fuel. Referring to an example above, it is seen that a 200-cal/g pulse would result in a cladding strain of more than 2.9 percent in high burnup fuel whereas that same pulse would result in a cladding strain of less than 1 percent in fresh fuel.

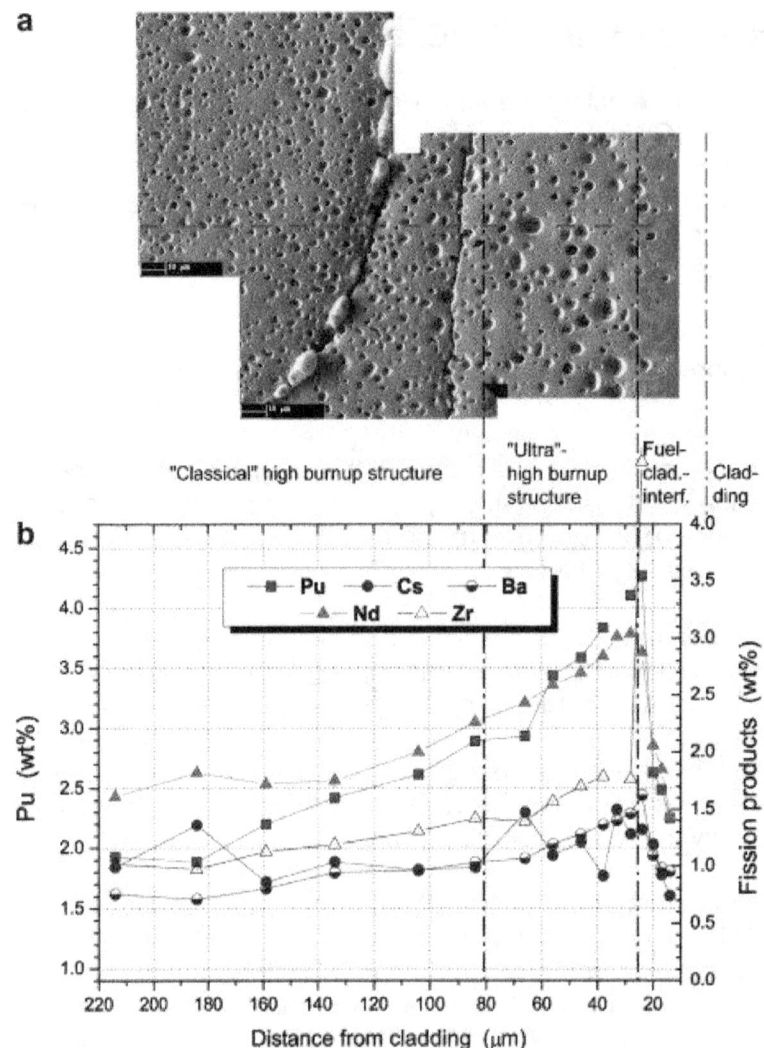

Figure 8 (a) Scanning electron microscope image showing the ultrahigh-burnup rim structure and (b) elemental distribution as a function of the distance from the fuel-to-cladding interface in a fuel rod with a burnup of about 105 GWd/t (Figure 1 in Ref. 8)

Because of the difficulties in calculating pellet expansion, gap closure, and cladding strain, empirical data can be used to obtain the strain values. Figure 11 shows data from RIA tests with high-burnup fuel in the CABRI test reactor in France and the Nuclear Safety Research Reactor (NSRR) in Japan (Ref. 2). In general, a trend from higher strain for PWR fuel with the largest operating pressure differentials (i.e., greatest cladding creepdown) to BWR fuel with modest operating pressure differentials to test reactor fuel with no operating pressure differential can be seen. The ordinate in Figure 11 shows only the measured plastic strain, so an additional elastic strain would exist in each test. Although the data exhibit a great deal of scatter, the total strain (plastic plus elastic) would clearly be greater than 2.9 percent for any of these cases at 200 cal/g in agreement with the above discussion. Although it appears that fresh cladding could easily accommodate a 1-percent strain during the power pulse considered, it is not clear that irradiated cladding would be so accommodating.

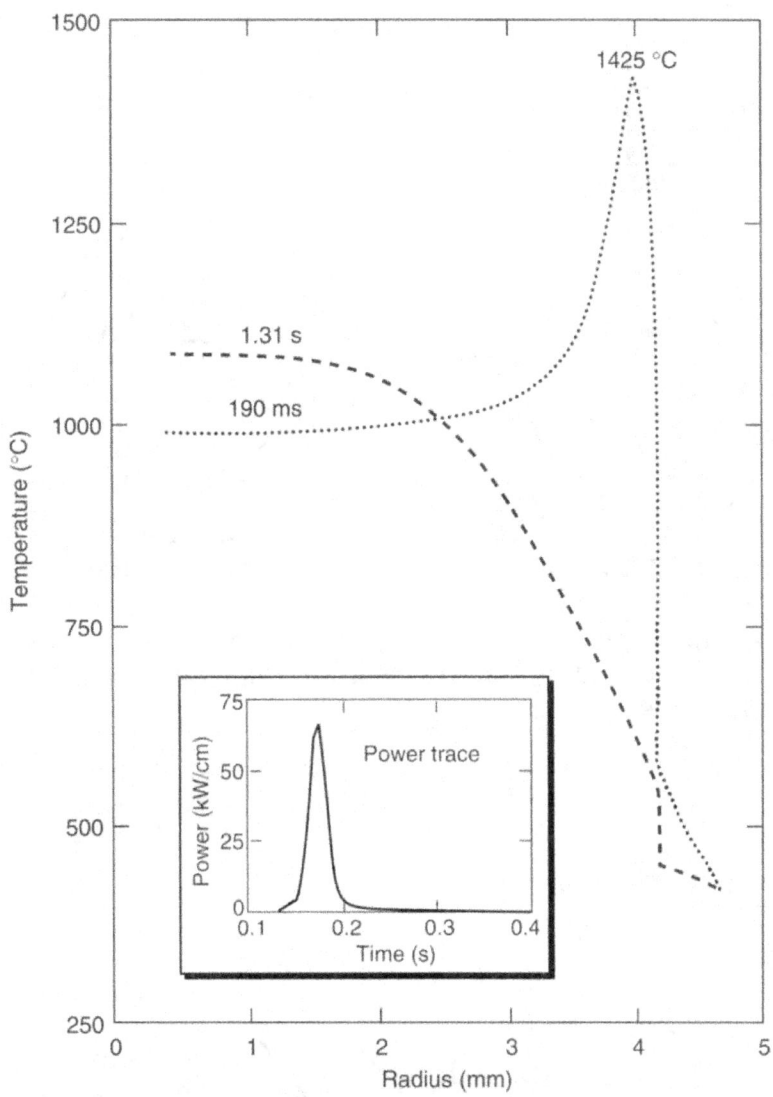

Figure 9 Edge-peaked pellet temperatures early in an RIA transient and center-peaked temperatures after significant heat transfer (based on Figure 24 and Figure 25 in Ref. 9)

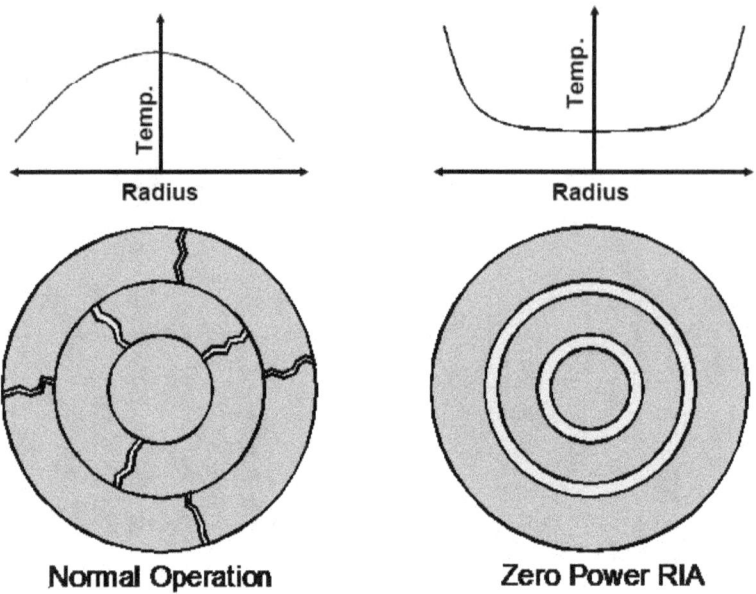

Figure 10 Thermal expansion algorithms for (a) normal operation (highest temperature at pellet centerline) and (b) a zero-power RIA (highest temperature at the edge) in high-burnup fuel

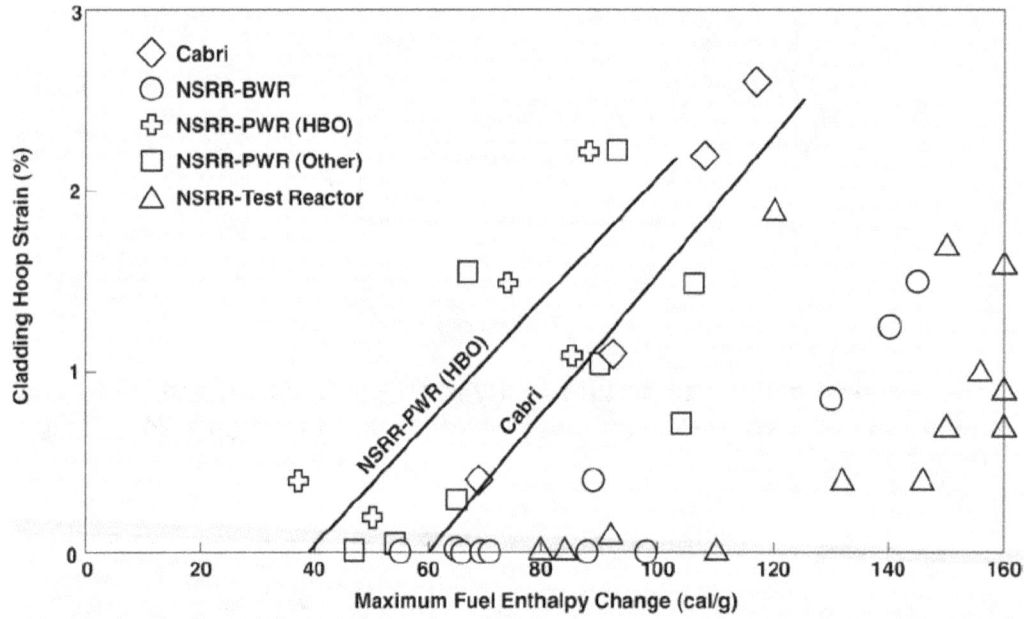

Figure 11 Plastic strain measured in nonfailed cladding as a function of maximum fuel enthalpy change for tests in the CABRI test reactor and the NSRR (Figure 7 in Ref. 2)

During normal operation, corrosion (i.e., slow oxidation) of the Zr-alloy cladding takes place according to the following reaction:

$$Zr + 2H_2O \rightarrow ZrO_2 + 2H_2$$

For each molecule of water (H2O) that reacts with the cladding to form zirconium dioxide (ZrO_2), two H atoms are liberated. Much of the H is swept away in the flowing coolant, but a fraction of the H is absorbed in the cladding because the monoclinic oxide structure is not very protective. At burnups around 60 GWd/t, typical cladding materials will contain H ranging from 100 to 800 weight-parts-per-million (wt.ppm), depending on the cladding alloy. Figure 12 shows that only about 50 wt.ppm of that H will be in solution at normal operating temperatures (Ref. 10). The remaining H will be in precipitates that will not dissolve during the nearly adiabatic period of an RIA power pulse. Like all precipitates, the hydrides will harden the cladding and reduce its ductility. The reduction in ductility will lead to cladding failure at lower and lower pulse energies as the H concentration increases (Figure 13). This type of failure, which is referred to as pellet-cladding mechanical interaction (PCMI) failure, generally starts with crack initiation in the brittle hydrided rim near the outside diameter (OD) and proceeds by a mixture of brittle and ductile failure modes through the rest of the cladding wall.

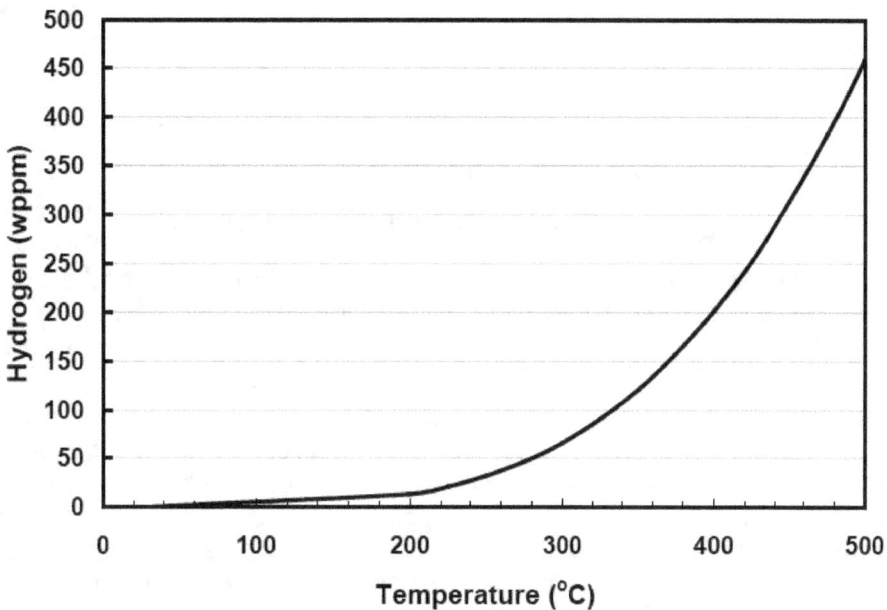

Figure 12 Solubility of H in Zr as a function of temperature (based on Ref. 10)

Two significant comments can be made about Figure 13. First, such data are usually evaluated as a function of maximum fuel enthalpy change. This change corresponds to the expansion during the transient and does not include any expansion that might take place slowly as the temperature increases from room temperature to operating temperatures (e.g., in a PWR). Irradiated cladding can usually accommodate expansion that takes place slowly by some combination of cladding creepout and realignment of pellet fragments. Second, at very low H concentrations, the cladding has enough ductility to survive the period of the power pulse, but the high-temperature mechanism described above for fresh fuel ultimately causes it to fail around 170 cal/g (total).

NUREG/KM-0004

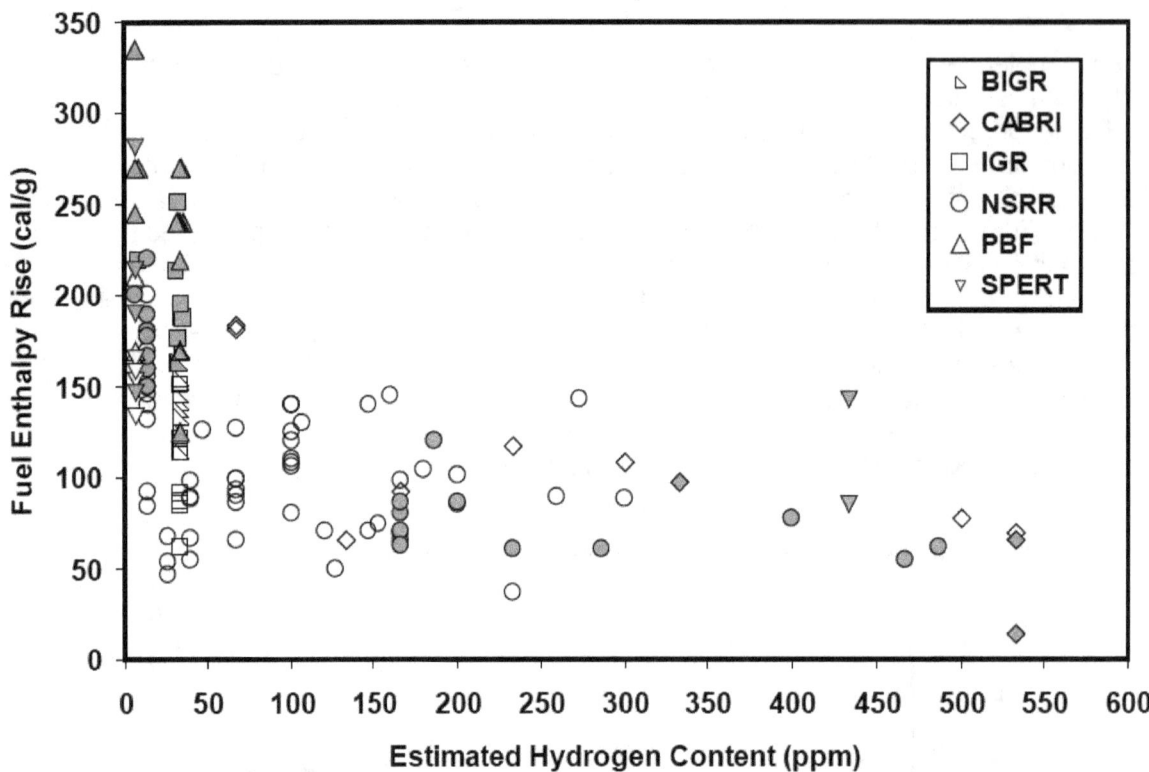

Figure 13 RIA test data for irradiated fuel, plotted as maximum fuel enthalpy change as a function of H in the cladding (H content was estimated when data were not available); solid symbols indicate cladding failure (based on Figure 3 of Ref. 2)

Finally, fuel dispersal that could lead to a steam explosion would occur by a totally different mechanism in high-burnup fuel than in fresh fuel. In fresh fuel, the fuel must melt to produce a sudden expansion that could expel hot fuel material. Fission gas bubbles, which accumulate on grain boundaries (shown as pores in Figure 8(a)), can cause such an expansion in high-burnup fuel. The rapid heating of fuel material, especially in the gassy ultrahigh-burnup rim region near the pellet periphery, causes the breakup of the fuel pellets and the entrainment of particles in gas that escapes through the cladding failure. Although data are sparse, most tests with high-burnup fuel in which the PCMI mechanism caused the failure experienced fuel dispersal that produced significant pressure pulses in the test apparatus. Based on the information presented in Figure 6, most fuel rods in a reactor would undergo a fuel enthalpy change of less than 70 cal/g during an RIA; however, Figure 13 shows PCMI failures in that range. Therefore, some irradiated fuel with high H content might experience cladding failure, fuel dispersal, and the potential for a steam explosion in an RIA if such fuel were located near high-worth control rods.

3. LOSS-OF-COOLANT ACCIDENTS

A pipe break or a leak in a component can initiate a LOCA. The LOCA at the Three Mile Island Nuclear Generating Station (Three Mile Island) in 1979 occurred when coolant was lost through a relief valve that did not close after a series of unplanned events. This accident evolved into a severe accident when an emergency core cooling pump, which had automatically started, was turned off because the operators mistakenly thought that the pressurizer volume was going "solid." The Three Mile Island accident offered many lessons, and improved instrumentation and procedures should significantly increase the likelihood that a future LOCA would be successfully terminated. This section will address LOCAs up to the point at which they are successfully terminated by an emergency core cooling system.

In BWRs and PWRs, a postulated LOCA begins with a blowdown of steam through the pipe break or leak as water evaporates rapidly (flashes). The increase in coolant voids shuts down the nuclear reaction, but heat stored in the fuel and decay heat from existing radionuclides must be removed to prevent excessive core temperatures. As the pressure in the vessel drops and core temperatures increase, emergency cooling systems are activated to add water to the vessel in an attempt to turn this temperature increase around. Figure 14 shows cladding temperatures and fuel rod pressures during a typical postulated LOCA that is successfully terminated by the emergency core cooling system (Ref. 11).

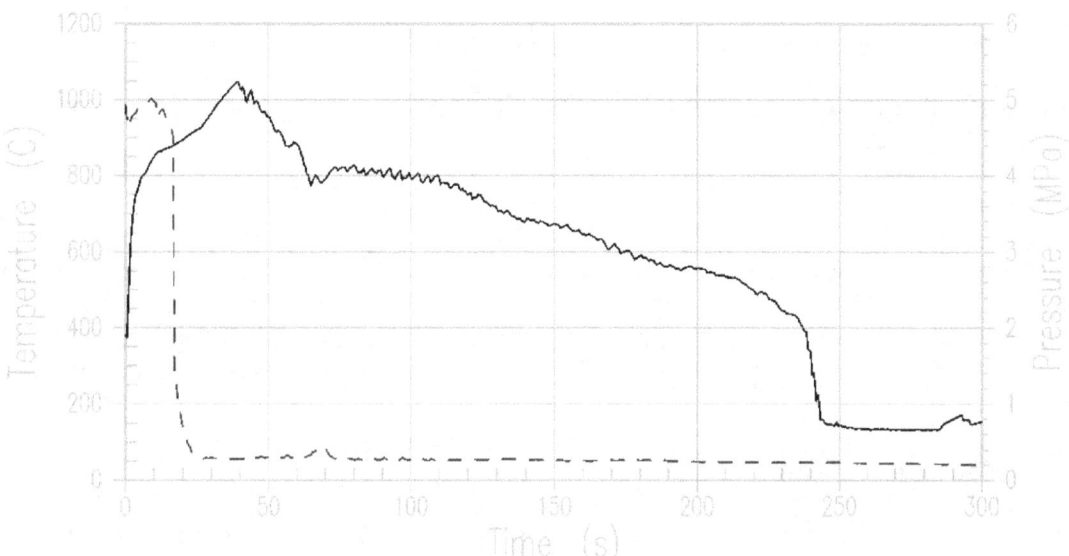

**Figure 14 Cladding temperature (solid line) and rod pressure (dashed line) during a
postulated LOCA (Figure 6 in Ref. 11)**

Emergency core cooling systems are designed to provide enough cooling to limit fuel damage, and the extent of predicted fuel damage in a safety analysis determines the adequacy of these systems. After a lengthy public hearing, the U.S. Atomic Energy Commission in 1973 concluded that fuel damage would be adequately limited if the cladding retained some ductility. Although limited ductility will not prevent fracturing, even marginal ductility will ensure that the cladding will exhibit its full tensile strength and be less susceptible to scratches and flaws than brittle material. This criterion was thought to be the best way to avoid fragmentation and keep fuel pellets within fuel rods in a coolable array. The ductility concept was adopted by most countries with nuclear power reactors, and cladding ductility under LOCA conditions has been extensively studied.

3.1 Loss-of-Coolant Accidents with Fresh Fuel

This discussion considers the events that take place as cladding temperatures rise and then fall during a typical postulated LOCA transient. Figure 15 illustrates these events, which the following sections discuss.

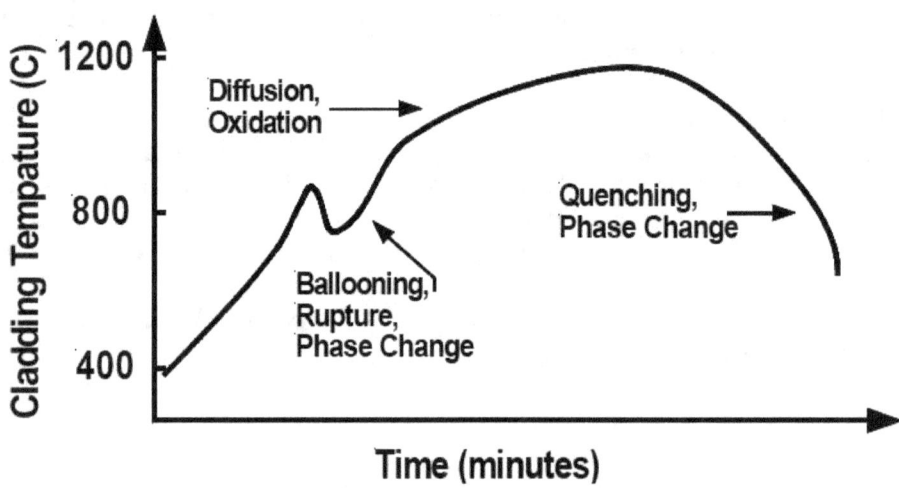

Figure 15 Illustration of temperature progression and associated phenomena during a postulated LOCA

3.1.1 Phase Change

The first event for Zr-alloy cladding (much of this discussion will assume Zircaloy-4) is a phase change from the alpha phase (hexagonal close-packed structure) to the beta phase (body-centered cubic structure) with the transition beginning at about 800 degrees Celsius and completed at about 975 degrees Celsius. The paragraphs below discuss subsequent phase transformations.

3.1.2 Ballooning and Rupture

Because fuel rods are pressurized with helium (He) to improve internal heat transfer, fuel rods have a high internal pressure when they are at operating temperatures. By design, these fuel rod pressures are limited in relation to system pressures (i.e., approximately 2,250 pounds per square inch (psi) in PWRs and approximately 1,035 psi in BWRs) to ensure that cladding does not creep away from the fuel pellets, thus degrading heat transfer. During a LOCA, system pressure is lost, and therefore the differential pressure across the cladding wall is large. Because of changes in its lattice structure, the cladding material softens enough to balloon and rupture in the same temperature range as the phase transformation. Figure 16 shows the temperatures and the corresponding engineering hoop stress at which rupture occurs for Zircaloy-4 (Ref. 12). The following equation relates the engineering hoop stress σ to the fuel rod pressure:

$$\sigma = d\ \Delta P\ /\ 2t,$$

where d is the cladding inside diameter (ID), t is the cladding thickness, and ΔP is the pressure differential.

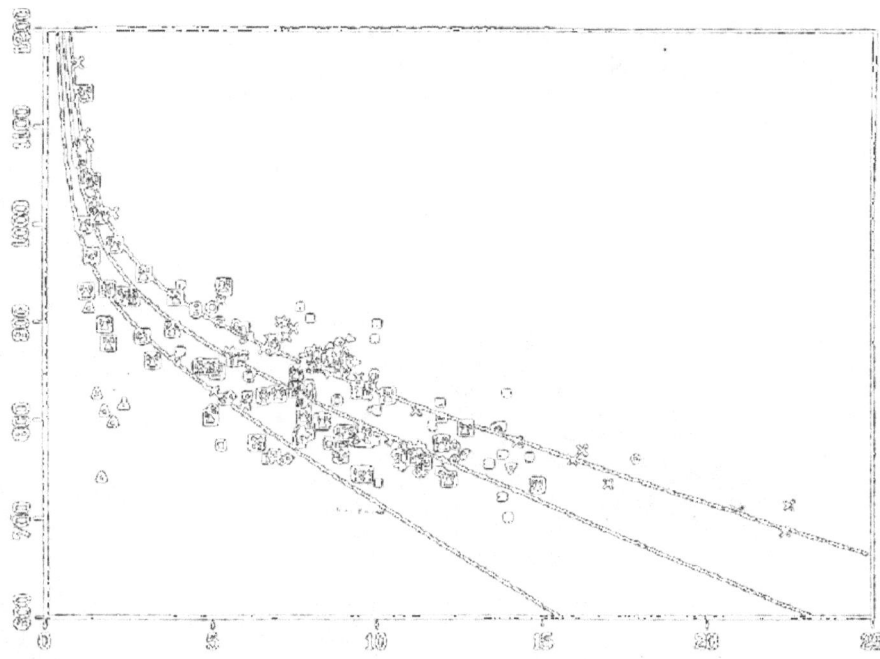

Figure 16 Correlations of rupture temperature and corresponding cladding stress for three heating rates (degrees Celsius per second) from data for Zircaloy-4 heated in steam (Figure 3 in Ref. 12)

Following a period of uniform outward deformation, a pressurized tube will eventually experience some local wall thinning that is triggered by a materials imperfection or, in the case of a fuel rod, a random local hot spot. As the wall thins, the local stress increases and the deformation will accelerate. This local deformation process is unstable, and the resulting balloon will soon burst. Because of the instability, it is not surprising that measured burst strains vary considerably from test to test; however, important trends are present. Figure 17 shows a large number of measured burst strains that reveal a maximum at around 800 degrees Celsius (Ref. 12). For this material (Zircaloy), 800 degrees Celsius marks the beginning of the phase transformation from alpha to beta. Thus, strains are apparently the largest when the material ruptures in the pure alpha phase, but the strains are reduced substantially when rupture occurs with a mixture of alpha and beta phases.

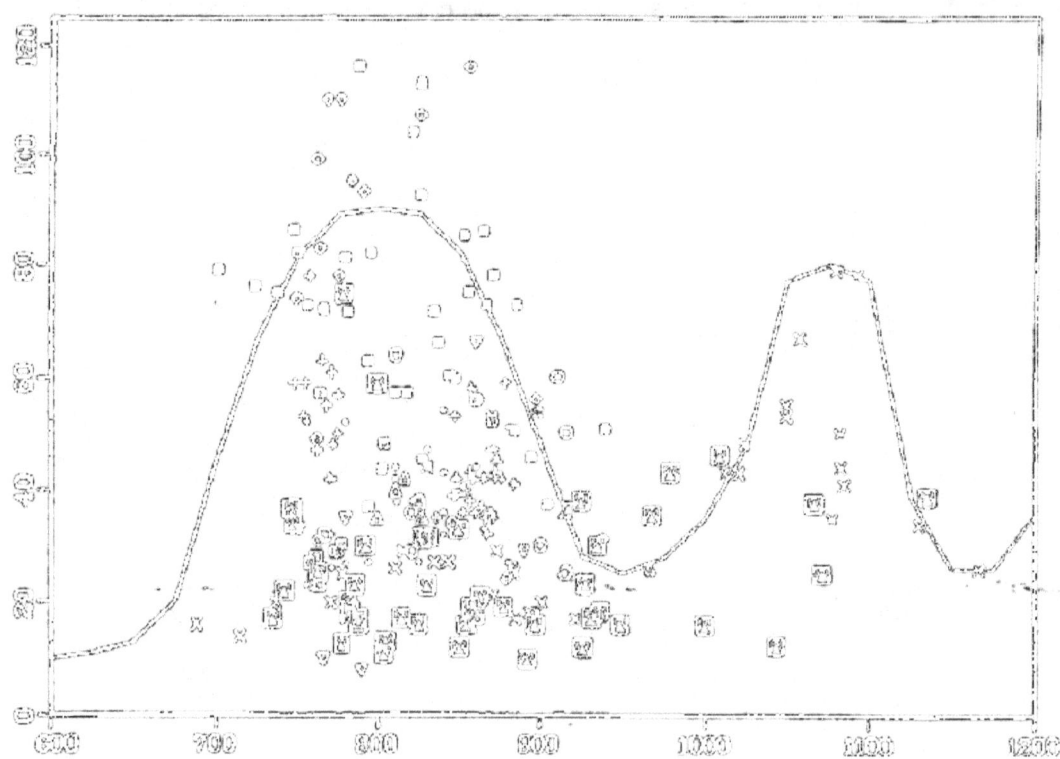

Figure 17 Maximum circumferential strain as a function of rupture temperature for Zircaloy cladding heated in steam (Figure 8 of Ref. 12)

Figure 18 shows an unirradiated fuel rod that has ruptured in a simulated LOCA transient (Ref. 13). The size of the cladding balloons at the time of rupture is important for two reasons. First, the swollen fuel rods reduce the flow area for coolant during the subsequent course of the LOCA. However, the increased surface area for cooling along with the turbulence introduced by the balloon usually result in a little extra cooling, rather than a temperature increase, in the balloon region.

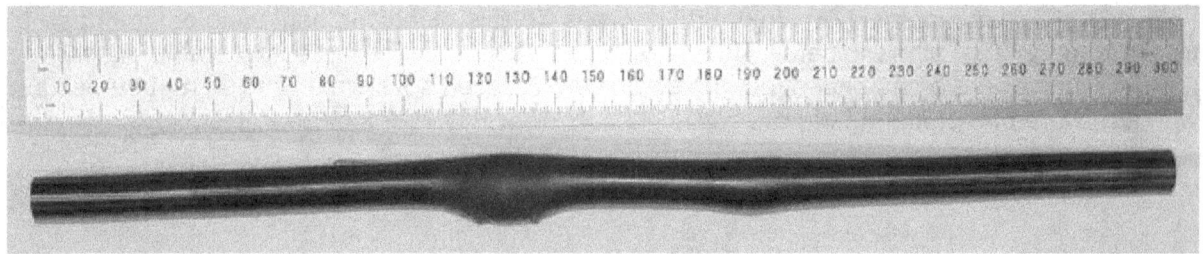

Figure 18 Ballooned and ruptured region of unirradiated Zircaloy-2 after undergoing LOCA conditions (Figure 217 in Ref. 13)

Second, cladding balloons will have thinner walls than those of the undeformed cladding, thus leading to higher oxidation when measured as a percentage of the wall thickness (equivalent cladding reacted (ECR)). In fact, oxidation will also take place on the inside of the cladding as well as on the outside because steam will enter through the rupture opening. This enhanced two-sided oxidation plays a significant role in most safety analyses.

3.1.3 Oxidation

When Zr is exposed to steam, it oxidizes at a more rapid rate than it does as a result of corrosion during normal operation. As the cladding temperature increases above 1,000 degrees C, the cladding surface oxidizes rapidly enough that more than 10 percent of the cladding wall thickness can be consumed during the period of a LOCA before emergency systems cool the core. The high-temperature tetragonal oxide structure is normally adherent and forms a protective layer on the cladding surface. Thus, H that is liberated in the high-temperature reaction does not get into the metal and is swept away with the flowing steam.

As the protective oxide builds up on the surface of the metal, it slows down the rate of oxidation by increasing the distance that oxygen has to diffuse to reach the metal surface. In the important temperature range of 1,000–1,500 degrees C, the amount of oxygen consumed per unit time is found experimentally to obey parabolic kinetics, as follows:

$$dw_g/dt = (k^2/2)/w_g,$$

where w_g is weight gain in grams per square centimeter of surface area, and k is a temperature-dependent coefficient. Under isothermal oxidation conditions, the integration of this equation is simply

$$w_g^2/2 = (k^2/2)\, t,$$

or

$$w_g = k\, t^{1/2},$$

which plots as a parabola.

Most chemical reactions and diffusion processes involve movement from one state or location to another that is separated by an energy barrier. Such reactions and processes accelerate rapidly as temperature increases according to the following equation, which was named after the Swedish chemist, Svante Arrhenius (1859–1927):

$$k = a\, \exp(-Q/RT),$$

where R is the universal gas constant; T is temperature in Kelvin; and the preexponential factor and activation energy, a and Q, respectively, are determined experimentally. Combining the parabolic equation with the Arrhenius equation gives the following equation:

$$w_g = a\, t^{1/2}\, \exp(-Q/RT)$$

Baker and Just (Ref. 14) performed an early study on Zr oxidation. Although it was found that the Baker-Just (B-J) correlation overestimates oxidation by as much as 30 percent around 1,200 degrees Celsius, this equation is provided below because it has been so widely used:

$$w_g\,(\text{B-J}) = 2.02\, t^{1/2}\, \exp(-22{,}750/RT)$$

Cathcart et al. (Ref. 15) performed a more recent and more accurate study of Zircaloy-4 oxidation. The Cathcart-Pawel (C-P) correlation is as follows:

$$w_g \text{ (C-P)} = 0.602 \, t^{1/2} \exp(-19{,}970/RT)$$

A related parameter that is often used is ECR, which is defined as the percentage of the cladding thickness that would be oxidized if all the oxygen stayed in the oxide layer as ZrO_2. This parameter is artificial because some of the oxygen diffuses into the metal, but it is useful and is directly related to weight gain by simple geometric factors along with factors based on the density of Zr (6,500 kilogram per cubic meter = 6.5 grams per cubic centimeter) and on the atomic masses of Zr and oxygen. The following equations provide the conversion for one-sided oxidation and two-sided oxidation:

one-sided oxidation ECR = $43.9 \, [(w_g/h)/(1 - h/d_o)]$, and

two-sided oxidation ECR = $87.8 \, w_g/h$,

where ECR is a percentage, h is cladding thickness in centimeters, and d_o is the cladding OD in centimeters.

The oxidation reaction of Zr in steam is exothermic with a reaction heat of 140.5 kilocalories per mole of Zr (6.45×10^6 J/kg) (Ref. 3). This reaction heat (also called metal-water reaction heat) can be greater than the decay heat from radioactive nuclides at higher temperatures, and the reaction heat will then dominate the heat-transfer situation. If dw/dt is the rate of Zr mass that reacted with steam, then the power P in kilowatts per centimeter generated by the metal-water reaction per centimeter length of cladding is as follows:

$$P = 20.3 \, d_o \, dw/dt$$

If all of the oxygen is assumed to form stoichiometric ZrO_2, then the mass of Zr that reacted is simply the following:

$$w = 91.2/32.0 \, w_g,$$

where w is in grams per square centimeter of surface area, 91.2 is the atomic mass of Zr, and 32.0 is the atomic mass of diatomic oxygen. Therefore, the following equation applies:

$$P = 57.9 \, d_o \, dw_g/dt$$

Figure 19 provides an example of linear power for various temperatures and various oxidation levels.

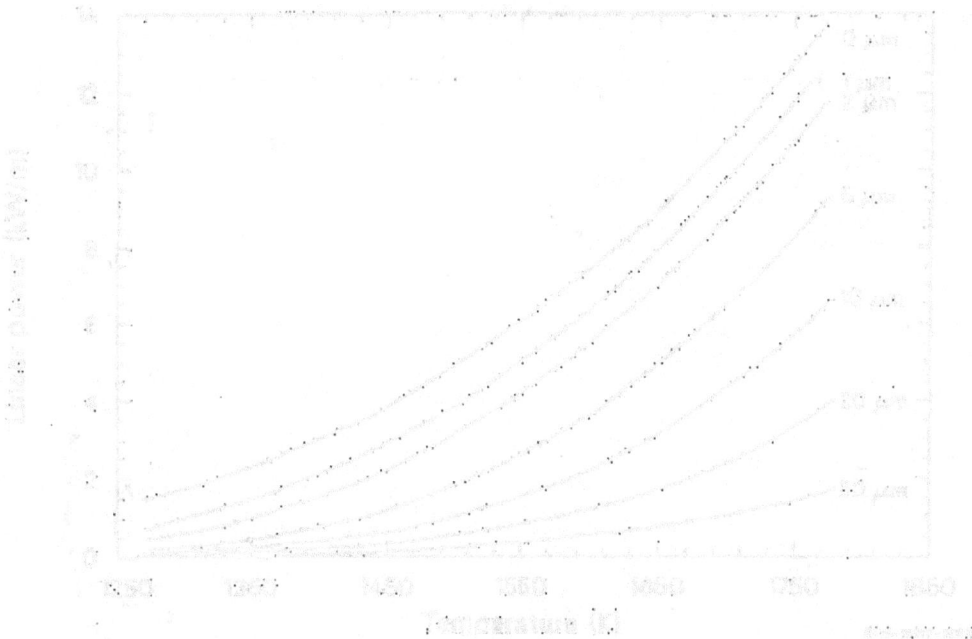

Figure 19 Linear power generation from the metal-water reaction for cladding with a diameter of 1.25 centimeters and various oxide thicknesses (Figure 4-69 in Ref.3)

3.1.4   Diffusion of Oxygen into the Metal

At the same time that oxidation is proceeding rapidly, oxygen from the cladding surface is diffusing rapidly into the cladding metal. However, Zr alloys in the beta phase cannot hold much oxygen. Therefore, as oxygen pours into the metal, near-surface regions with high oxygen concentrations change phase again—this time back to the alpha phase, which oxygen stabilizes (i.e., the so-called oxygen-stabilized alpha phase). The oxygen solubility limit in the alpha and beta phases depends on temperature, as shown in Figure 20 (Ref. 16). For example, at 1,200 degrees C, the beta phase can hold only about 0.7 weight-percent oxygen.

Finally, the emergency core cooling system will reflood the core, temperatures will begin to come down, and all of the cladding metal will return to the alpha phase. The imprint of the high-temperature phases can actually be seen microscopically at room temperature as shown in Figure 21 (Ref. 13). When the cladding temperature reaches its wetting (Leidenfrost) temperature in the range of 600–800 degrees C, the cladding will quench, and the temperature will fall quickly to the average water temperature. Because the containment is sealed, the containment pressure will rise somewhat, thus elevating the boiling point of water to about 135 degrees C; therefore, the cladding does not return exactly to room temperature. At this lower temperature, oxygen diffusion is effectively stopped and the high-temperature oxygen distribution is frozen into the cladding, as illustrated in Figure 22.

NUREG/KM-0004

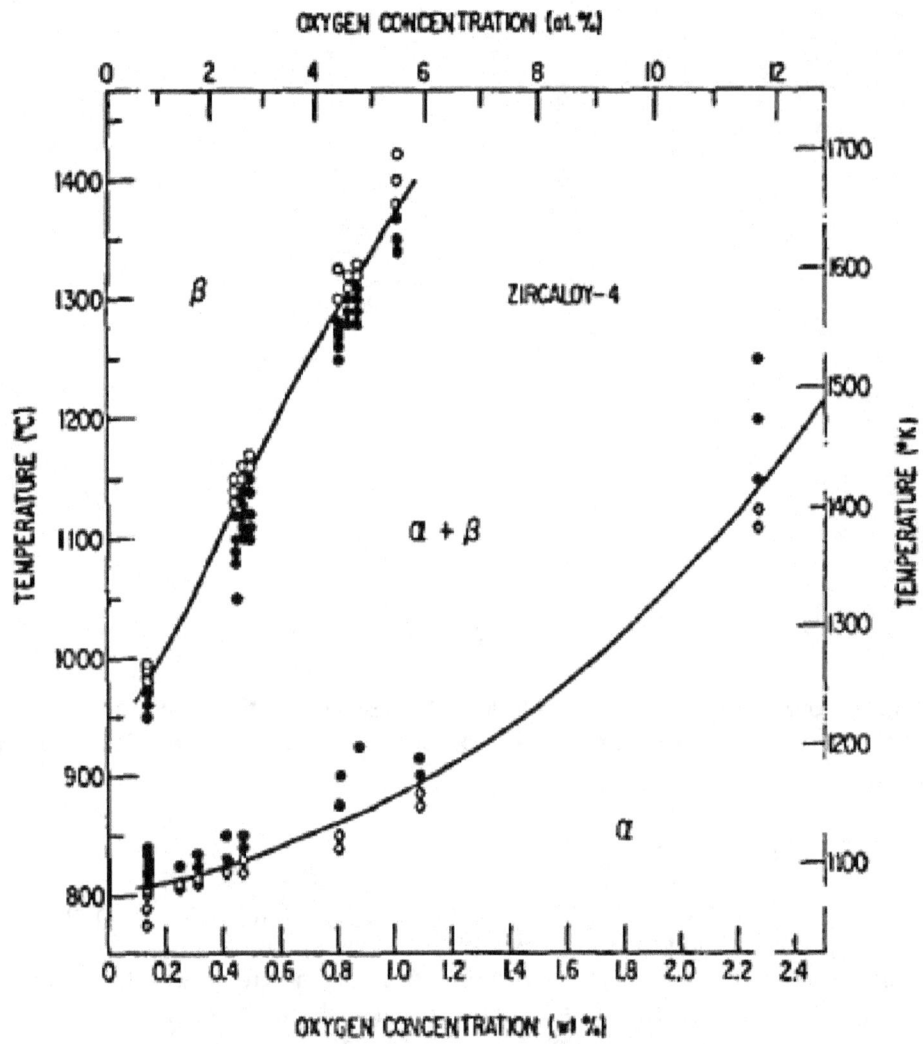

Figure 20 Pseudobinary Zircaloy-oxygen phase diagram (Figure 4 in Ref. 16)

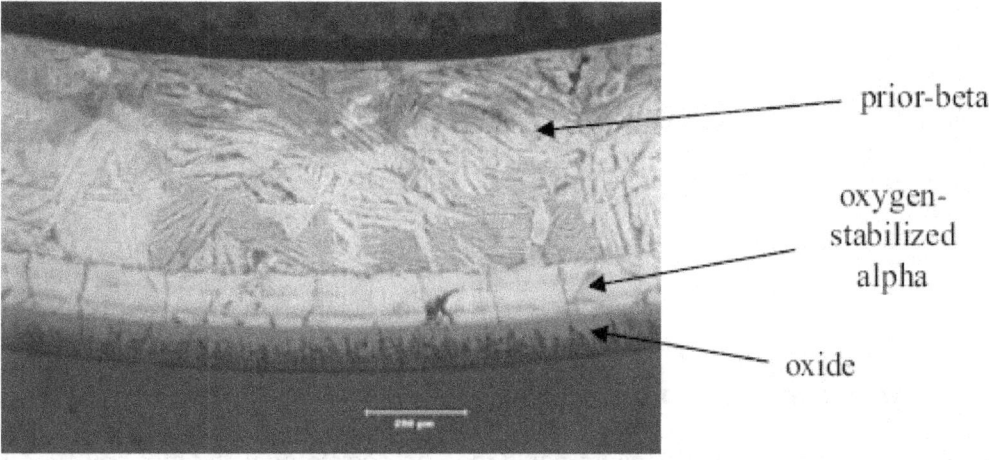

Figure 21 Microscopic image of unirradiated Zircaloy-2 after oxidation in steam at 1,200 degrees Celsius for 600 seconds at room temperature (Figure 4 in Ref.13)

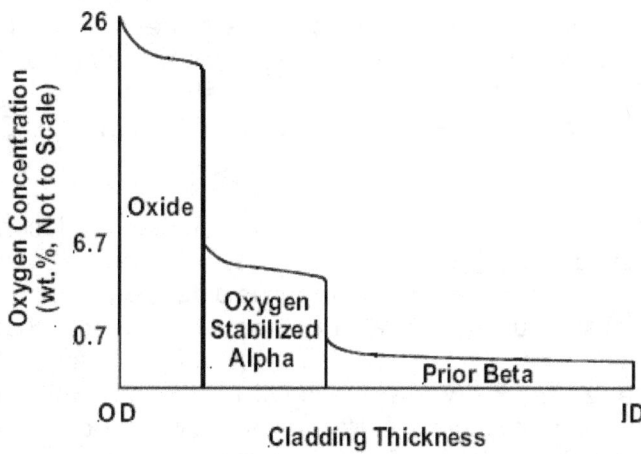

Figure 22 Qualitative diagram of oxygen concentration in Zircaloy cladding exposed to steam at high temperature

Zircaloy metal containing more than about 0.6 wt-percent oxygen is brittle. Therefore, only the low-oxygen material, which was in the beta phase when it was at high temperature, is ductile. To determine the LOCA conditions under which fuel rod cladding becomes brittle, mechanical tests are usually performed at 135 degrees C on short lengths, or rings, of cladding specimens that have been exposed to high-temperature steam for different lengths of time. Figure 23 shows a ring-compression test that is generally used for this purpose. During compression, maximum tensile stresses are generated at the 12 o'clock and 6 o'clock positions on the ID and at the 9 o'clock and 3 o'clock positions on the OD. Zero ductility is assumed when failure occurs at one of these locations with a very small (1 percent) permanent strain.

NUREG/KM-0004

Figure 23 Diagram of ring-compression test

The transition from ductile to brittle failure could then be correlated with time; however, the time would strongly depend on the steam temperature. Instead, the ductile-to-brittle transition is usually correlated with the calculated (not measured) ECR using either the B-J equation or the C-P equation. This is in effect a variable transformation that combines both time and temperature. Using this method, the transition from ductile-to-brittle behavior is found at approximately 17-percent ECR as calculated with the C-P correlation for modern Zr-alloy cladding materials; this result is nearly independent of temperature. In other words, enough oxygen diffuses into the metal to cause embrittlement in the same amount of time that it takes Zircaloy to oxidize to 17-percent ECR. Because this is just a mathematical transformation, the same Zircaloy oxidation equation is used to evaluate embrittlement in all Zr-alloy cladding materials.

Interestingly, although the diffusion of oxygen controls the rate of oxidation and embrittlement of the cladding materials, the diffusion mechanisms are very different. In one case, oxygen is diffusing through the oxide; in the other case, oxygen is diffusing through the metal. Oxides such as ZrO_2 have crystal structures that are ionic in nature, and local charge neutrality must be maintained. Therefore, if an impurity such as niobium (Nb) with a valence of +5 gets into the oxide, which normally has Zr cations with a valence of +4, the local lattice defect concentration is modified, and the diffusion rate is altered. Impurities like Nb are called aliovalent impurities. Note that tin (Sn) has a valence of +4 just like Zr; therefore, Sn impurities do not significantly affect diffusion in ZrO_2. Consequently, alloys like M5 (Zr-1 percent Nb) oxidize at different rates than those of Zircaloy (Zr-1.4 percent Sn).

On the other hand, diffusion of oxygen in Zr metal results from an interstitial mechanism. The number of interstitial locations is fixed and not affected by substitutional impurities such as Sn and Nb. Therefore, the rate of embrittlement of different Zr-based cladding alloys should not be sensitive to alloy composition. Figure 24 identifies these diffusion mechanisms and indicates the oxygen solubility limits in the different phases.

Figure 24 Illustration of diffusion mechanisms in Zr-based cladding alloys

3.1.5 Hydrogen Absorption

Two special circumstances exist in which H can be absorbed in the cladding metal during a LOCA transient. One involves a phenomenon called breakaway oxidation, and the other involves ruptured balloons. The significance of H absorption is that H acts like a catalyst and accelerates the embrittlement process partly by increasing the rate of diffusion of oxygen in the metal.

As stated previously, the reaction of Zr with steam during a LOCA normally produces a protective oxide layer on the cladding surface; however, this is not always the case. The normal protective tetragonal form can transform to a monoclinic form under some circumstances. The monoclinic form of ZrO_2 is fragile and contains many cracks that give H access to the metal. Very soon after the monoclinic oxide forms at LOCA temperatures, sufficient H enters the metal to cause embrittlement long before the 17-percent ECR level is reached.

Several conditions can lead to early breakaway oxidation (Refs. 13 and 17). A rough surface on the cladding promotes the formation of a monoclinic oxide by providing alternate compressive and tensile stress states for crystal growth. Fluorine, which has been used in some etchants, also promotes breakaway oxidation. Impurities in Zr produced by the Kroll process seem to be beneficial in preventing breakaway oxidation compared with pure electrolytic Zr. However, the origins of breakaway oxidation are not well understood, and other factors may be involved.

Figure 25 shows two cladding materials that have the same nominal composition but were fabricated differently. The older Russian E110 alloy experienced breakaway oxidation in a few hundred seconds compared with the French M5 alloy, which shows no evidence of breakaway oxidation after 2,400 seconds. After learning about the factors mentioned above, the Russian manufacturer, TVEL, altered the manufacturing process to avoid early breakaway oxidation in the E110 alloy (Ref. 18).

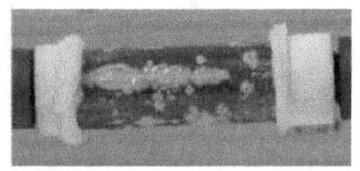

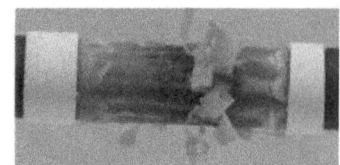

E110, 290 sec   E110, 1400 sec

Note that E110 and M5
are nominally the same
alloy (Zr-1%Nb)

M5, 2400 sec

**Figure 25 Older E110 cladding showing early breakaway oxidation at 1,000 degrees C
compared with modern M5 cladding, which is resistant to breakaway oxidation
(based on Figure 88 in Ref. 13 and on related unpublished figures)**

3.2   Loss-of-Coolant Accidents with High-Burnup Fuel

The same properties of high-burnup fuel that alter RIA behavior (Section 2.2) also alter LOCA
behavior but in an entirely different manner.

3.2.1   Effect of Hydrogen

All of the H that existed as hydride precipitates in high-burnup fuel during normal operation will
go into solution at LOCA temperatures, as shown in Figure 12 (off scale). As mentioned above,
H in the cladding metal increases the rate of oxygen diffusion at LOCA temperatures and
accelerates the embrittlement process so that embrittlement occurs in less time than it takes to
oxidize Zircaloy to 17-percent ECR. Figure 26 shows this effect for cladding with various H
concentrations, using the calculated ECR (C-P correlation) as the measure of time at
temperature. Data points in this figure that are not labeled with burnup values correspond to
unirradiated cladding samples that were either as-fabricated or artificially charged with H. Each
point in this figure represents a series of ring-compression tests that were conducted to locate
the embrittlement threshold. The trend apparently applies equally to all of the Zr alloys, as
expected (see Figure 24 and related text).

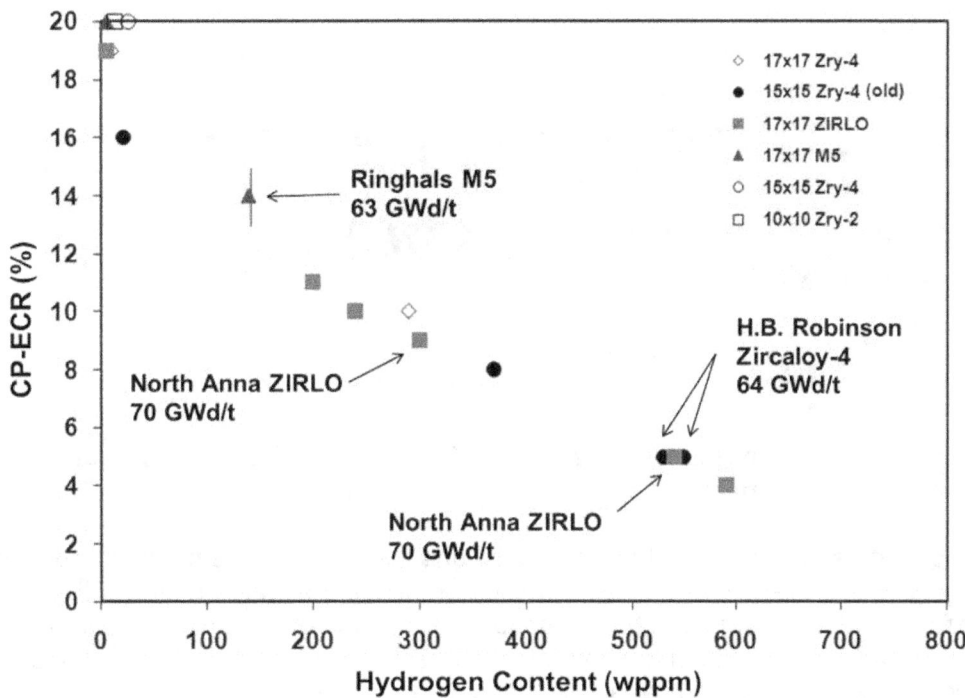

Figure 26 Embrittlement threshold for cladding specimens exposed to steam at 1,200 degrees Celsius and quenched after slow cooling to 800 degrees Celsius (Ref. 19)

3.2.2 Inside Diameter Oxygen Pickup

The pellet-to-cladding gap closes early during the burnup process, and long-term contact between the pellet and cladding at operating temperatures results in diffusion welding, or bonding. This process occurs earlier in PWRs than it does in BWRs because of the different design pressures and cladding thicknesses, and the bonding is fully developed by about 50 GWd/t in PWRs and 60 GWd/t in BWRs. Bonding can be seen microscopically (Figure 27), and technicians who remove fuel from high-burnup specimens for testing purposes have found tangible evidence of this bonding. When bonding is present, impact drilling is usually required to remove the fuel, and the bonding interaction layer along with attached pieces of fuel remain attached to the cladding.

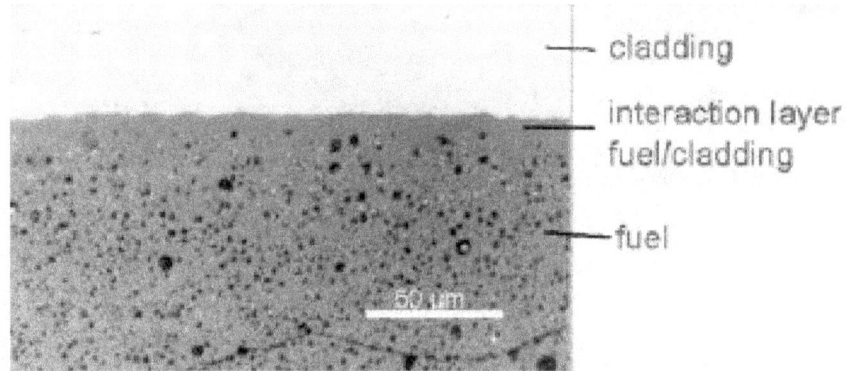

Figure 27 Interaction bonding layer between UO2 fuel and Zircaloy-4 cladding in a high-burnup specimen at 83 GWd/t (Figure 10 in Ref. 20)

The bonding layer is mostly ZrO_2. Figure 28 provides insights into this bonding process. This Ellingham plot shows the free energy of formation of oxides with various metals that are relevant to nuclear fuel materials. Note that the formation energy is nearly the same for Zr, uranium, and Pu (prevalent in high-burnup fuel) and that all of these metals form dioxides. Thus, oxygen can be exchanged freely among them when the cladding and the fuel are in contact.

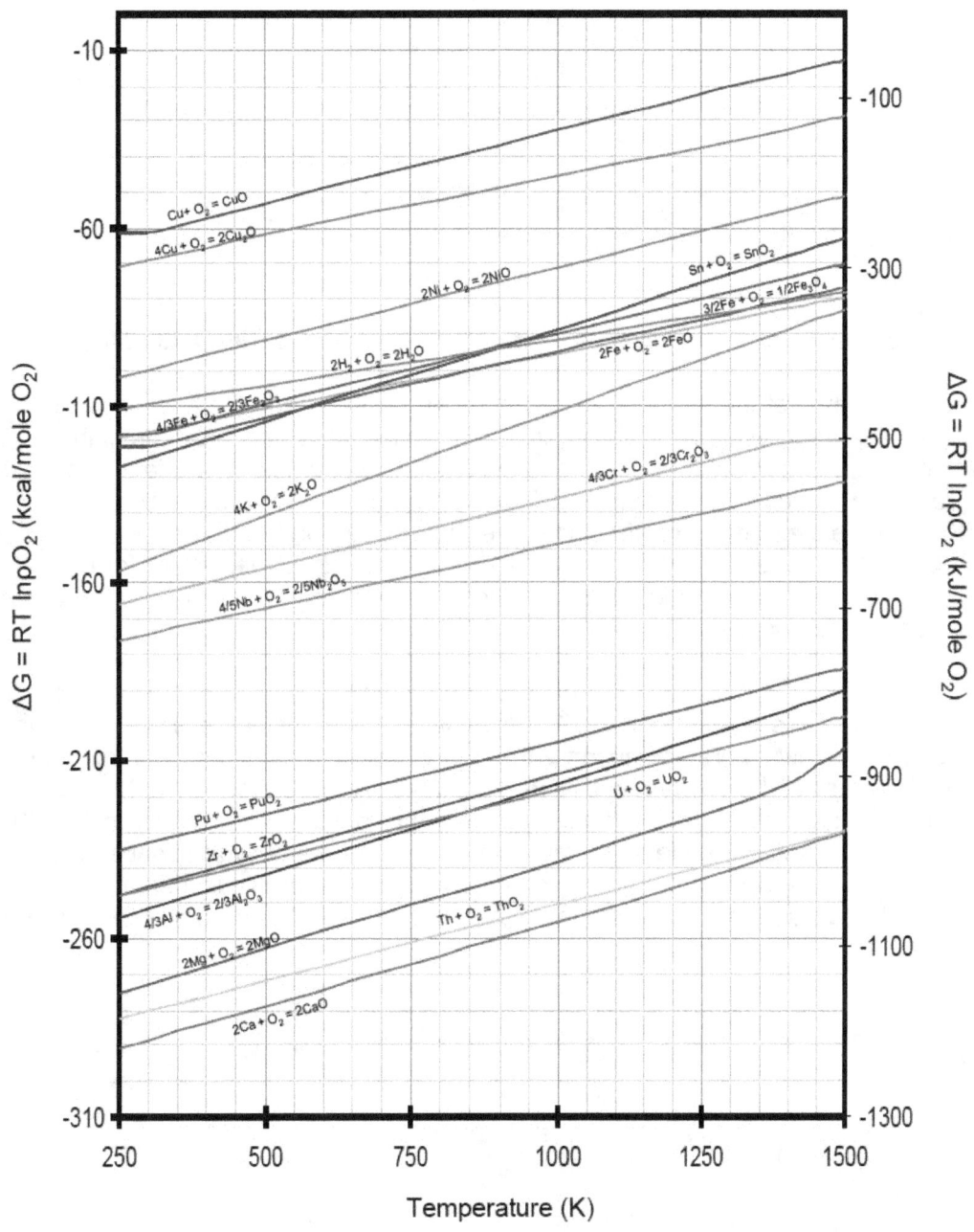

Figure 28 Free energy of formation of oxides of Zr-alloy constituents and some impurities (Figure 1 in Ref. 13)

The consequence of fuel-to-cladding bonding for LOCA behavior is that an oxygen source is then present on the cladding ID and on the OD, as indicated in Figure 29.

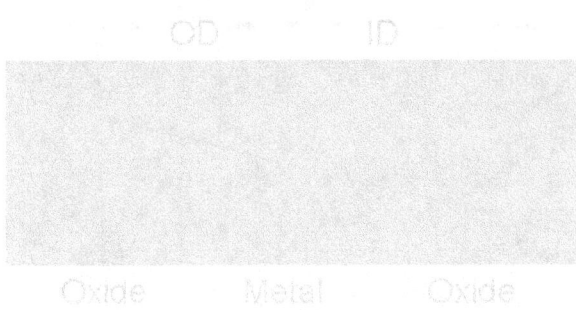

Figure 29 Oxygen sources for diffusion into cladding metal during a LOCA

During the high-temperature portion of a LOCA, cooling is poor, and the heat flux through the cladding is small. As a result, the cladding is nearly isothermal when it is at its highest temperatures and when most of the oxygen diffusion is taking place. Therefore, the oxygen diffusion rate is the same on the ID and the OD. Furthermore, because the quantity of oxygen needed to embrittle the metal is rather small (see Figures 22 and 24), the thicknesses of the OD and ID oxide layers are relatively unimportant—usually enough oxygen exists for embrittlement. Thus, when a well-developed bonding layer is present, the same amount of oxygen should enter the metal from the ID and the OD.

Figure 30 provides confirmation of this process. These micrographs were taken on a high-burnup fuel rod after testing under LOCA conditions in the Halden boiling heavy water reactor (Figure 27 shows a pretest micrograph). In Figure 30, the microscopist measured and labeled the thickness of the oxygen-stabilized alpha layer at the OD and ID at two azimuthal locations. In all cases, this alpha layer had the same thickness (about 20 microns) on the OD and ID. Because the interior edge of the alpha layer is fixed at an oxygen concentration of about 6.7 wt-percent (see Figure 22), the oxygen concentration profile is the same on the ID and OD, thus demonstrating that the same amount of oxygen entered from both directions.

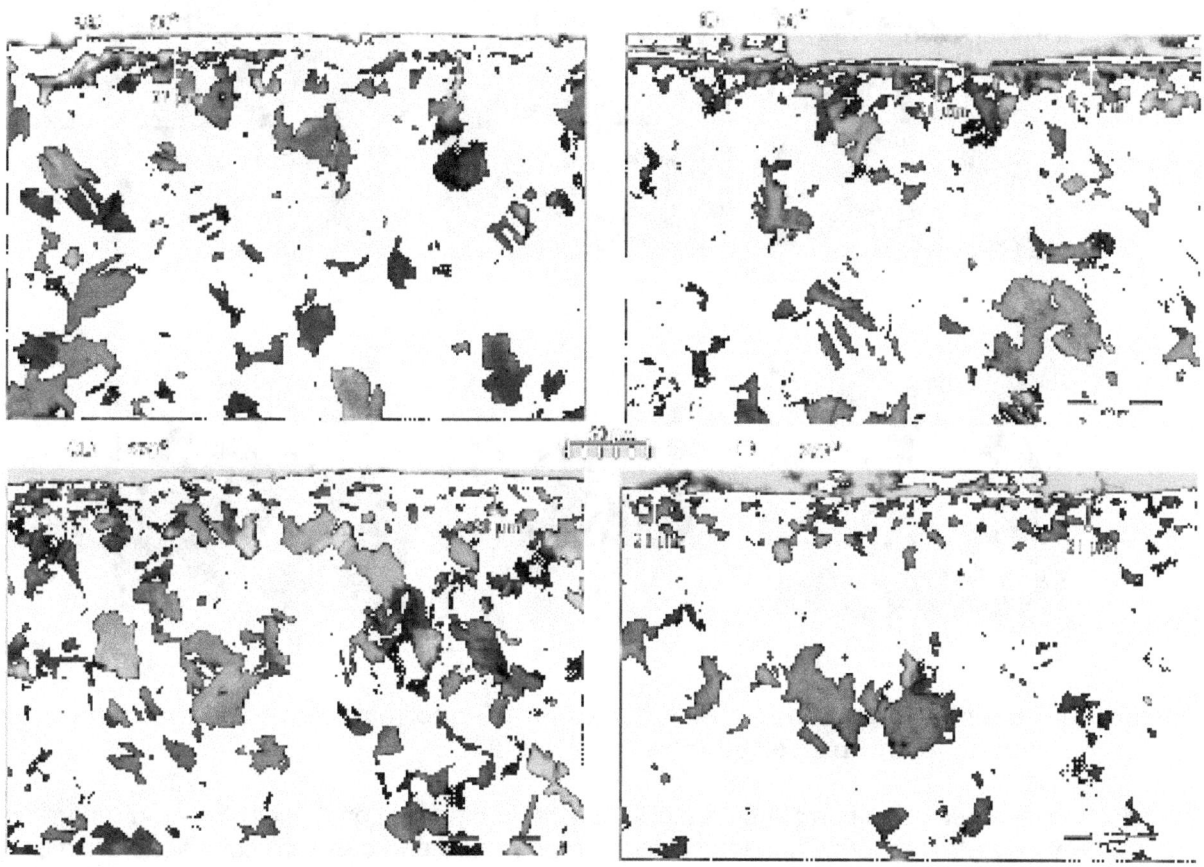

**Figure 30 Grain morphology and alpha layer at the OD and ID of a high-burnup fuel rod
exposed to LOCA conditions (Supplementary Figure 3.5.24 in Ref. 20)**

3.2.3 Fuel Relocation

Because UO_2 is a brittle ceramic, reactor startup and subsequent power changes cause fuel
pellets to crack even at moderately low burnups. Figure 31 shows pellet fragments in two such
irradiated fuel rods that have been exposed only to normal operating conditions. If these pellet
fragments could move axially into the enlarged volume of a cladding balloon during a LOCA, the
heat generation rate within the balloon caused by radionuclide decay and stored heat would
increase. This temperature increase would accelerate the embrittlement process in the balloon
region.

2 500 MWd/t   C 6

35 000 MWd/t   G 1.6

Figure 31 Fuel pellet fragments in PWR fuel rods after normal operation to low and medium burnup levels (Figure 30 in Ref. 21)

As cladding temperatures rise during a LOCA, unstable deformation, ballooning, and rupture will all occur within a few seconds once the rupture temperature and pressure conditions are met (Figure 16). Immediately thereafter, cracked pellet fragments from just above the ballooned region move into the enlarged volume of the balloon. Gravity drives this movement, or relocation. Figure 32 shows an example of this process.

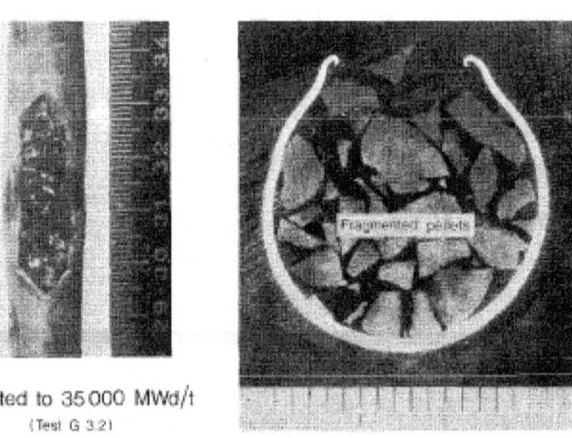

Irradiated to 35 000 MWd/t
(Test G 3.2)

Figure 32 Cross-section of the rupture region of an irradiated fuel rod after testing under LOCA conditions (Figure 13 in Ref. 21)

Randomly oriented granules in a rubble bed or heap are characterized by a packing fraction that is less than unity. Figure 32 clearly shows that the fragmented fuel particles that have relocated into the balloon have a lower overall density than that of the original fuel pellets from which the fragments came. Therefore, a certain minimum balloon size would be required to realize an actual increase in fuel mass within the balloon region. Figure 33 shows that the relative volume increase needed to accommodate additional mass in the balloon is about 18 percent for a large number of simulated LOCA tests with irradiated fuel rods. This corresponds to a packing fraction of about 82 percent on average. A diametral strain of less than 9 percent will give a volume increase of 18 percent; therefore, Figure 18 clearly shows that essentially all balloons that occur during a LOCA will be large enough to accommodate extra mass by fuel relocation.

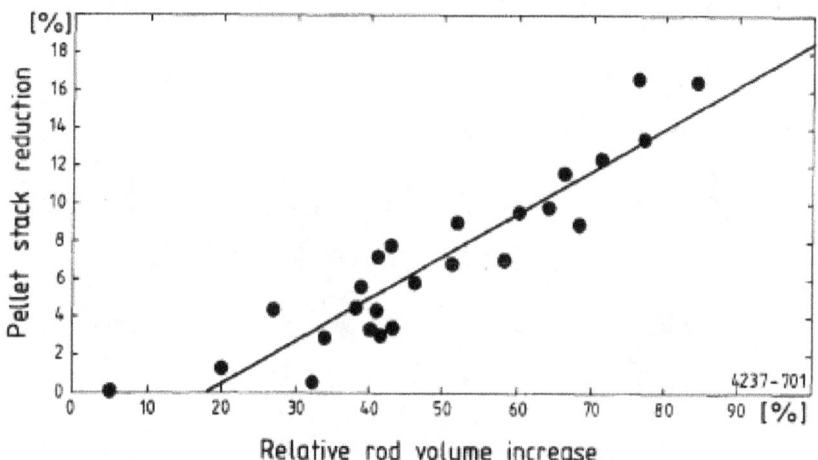

Figure 33 Pellet stack reduction as a function of the increase in volume in cladding balloons for preirradiated fuel rods (Figure 32 in Ref. 21)

It should be noted here that Figure 8(a) shown in a previous section displays a special case of fuel relocation that can occur at very high burnups. Porosity is very high in the rim region (22 percent in the case shown), and the pores are filled with fission gas at high pressure. When the fuel rod pressure is released at the time of cladding rupture, the unopposed pressure within the pores fractures the brittle fuel material, and very small fuel particles are entrained in the fission gas and fill gas that escape through the rupture opening.

Figure 34 shows the gamma scan of a fuel rod segment that had been subjected to LOCA conditions in the Halden boiling heavy water reactor. This fuel had a burnup of about 92 GWd/t, and more than 40 percent of the fuel volume had the rim structure. In this test, the entire upper portion (19 centimeters) of the fuel above the rupture opening was lost during rod depressurization. Researchers at Argonne National Laboratory have run similar out-of-pile tests on fuel that had a burnup of about 57 GWd/t with the observed loss of fuel amounting to less than that in one fuel pellet. Earlier LOCA tests at the Karlsruhe Nuclear Research Center on fuel with burnups no greater than 35 GWd/t resulted in no loss of fuel. Thus, very high burnup fuel with a well-developed rim structure may be susceptible to significant loss of fuel particles during a LOCA transient.

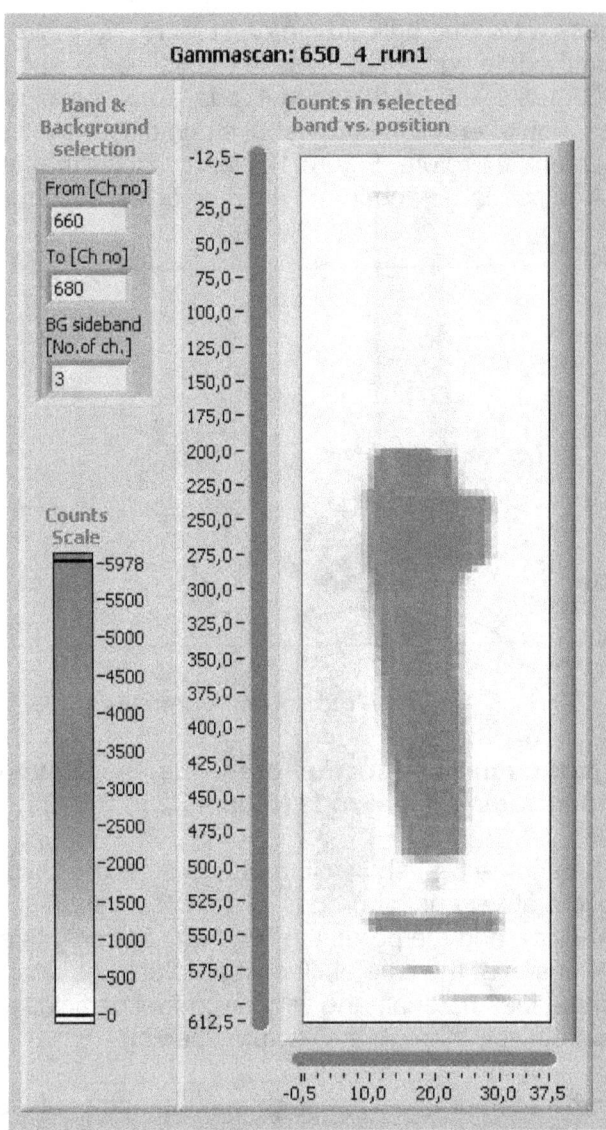

Figure 34 Gamma scan of very-high-burnup (approximately 92 GWd/t) fuel rod showing major loss of fuel material after LOCA testing (Figure 15 in Ref. 22)

4. REFERENCES[1]

1. Sains, A., "Forsmark-3, Oskarshamn-3 Face Outages for Control Rod Cracking," *Nucleonics Week*, 49:14–15, October 30, 2008.

2. Meyer, R.O., "An Assessment of Fuel Damage in Postulated Reactivity-Initiated Accidents," *Nuclear Technology*, 155:294–311, September 2006.

3. Siefken, L.J., E.W. Coryell, E.A. Harvego, and J.K. Hohorst, "SCDAP/RELAP5/MOD 3.3 Code Manual: MATPRO—A Library of Materials Properties for Light-Water-Reactor Accident Analysis," NUREG/CR-6150, Vol. 4, Rev. 2, January 2001, Agencywide Documents Access and Management System (ADAMS) Accession No. ML010330424.

4. Geelhood, K.J., C.E. Beyer, and M.E. Cunningham, "Modifications to FRAPTRAN To Predict Fuel Rod Failures Due to PCMI during RIA-Type Accidents," *Proceedings of the 2004 International Meeting on Light Water Reactor Fuel Performance*, Orlando, FL, American Nuclear Society, pp. 585–595, 2004.

5. Seiffert, S.L., Z.R. Martinson, and S.K. Fukuda, "Reactivity Initiated Accident Test Series, Test RIA-I-1 (Radial Average Fuel Enthalpy of 285 cal/g) Fuel Behavior Report," NUREG/CR-1465, September 1980.

6. Meyer, R.O., R.K. McCardell, H.M. Chung, D.J. Diamond, and H.H. Scott, "A Regulatory Assessment of Test Data for Reactivity-Initiated Accidents," *Nuclear Safety*, 37:271–288, October–December 1996.

7. MacDonald, P.E., S.L. Seiffert, Z.R. Martinson, R.K. McCardell, D.E. Owen, and S.K. Fukuda, "Assessment of Light-Water-Reactor Fuel Damage during a Reactivity-Initiated Accident," *Nuclear Safety*, 21:582–602, 1980.

8. Romano, A., M.I. Horvath, and R. Restani, "Evolution of Porosity in the High-Burnup Fuel Structure," *Journal of Nuclear Materials*, 361:62–68, March 2007.

9. Papin, J., M. Balourdet, F. Lemoine, F. Lamare, J.M. Frizonnet, and F. Schmitz, "French Studies on High-Burn-up Fuel Transient Behavior under RIA Conditions," *Nuclear Safety*, 37:289–327, October–December 1996.

10. McMinn, A., E.C. Darby, and J.S. Schofield, "The Terminal Solid Solubility of Hydrogen in Zirconium Alloys," *12th International Symposium on Zirconium in the Nuclear Industry*, Toronto, Ontario, American Society for Testing and Materials STP-1354, pp. 173–195, 2000.

11. Nissley, M.E., C. Frepoli, and K. Ohkawa, "Realistic Assessment of Fuel Rod Behavior under Large-Break LOCA Conditions," *Proceedings of the Nuclear Fuels Sessions of the 2004 Nuclear Safety Research Conference*, NUREG/CP-0192, pp. 231–273, October 2005, ADAMS Accession No. ML052980524.

---

[1] Some earlier documents have been made available in the Electronic Reading Room at www.nrc.gov under ADAMS Documents. For those documents, the ADAMS accession number is provided in the reference list. More recent NRC documents can also be found at www.nrc.gov.

12.   Powers, D.A., and R.O. Meyer, "Cladding Swelling and Rupture Models for LOCA Analysis," NUREG-0630, April 30, 1980, ADAMS Accession No. ML053490337.

13.   Billone, M., Y. Yan, T. Burtseva, and R. Daum, "Cladding Embrittlement during Postulated Loss-of-Coolant Accidents," NUREG/CR-6967, July 2008, ADAMS Accession No. ML082130389.

14.   Baker, L., and L.C. Just, "Studies of Metal-Water Reactions at High Temperatures. III. Experimental and Theoretical Studies of the Zirconium-Water Reaction," Argonne National Laboratory (ANL)-6548, May 1962, ADAMS Accession No. ML050550198.

15.   Cathcart, J.V., R.E. Pawel, R.A. McKee, R.E. Druschel, G.J. Yurek, J.J. Campbell, and S.H. Jury, "Zirconium Metal-Water Oxidation Kinetics IV. Reaction Rate Studies," Oak Ridge National Laboratory (ORNL)/NUREG-17, August 1977, ADAMS Accession No. ML052230079.

16.   Chung, H.M., and T.F. Kassner, "Embrittlement Criteria for Zircaloy Fuel Cladding Applicable to Accident Situations in Light-Water Reactors: Summary Report," NUREG/CR-1344, January 1980, ADAMS Accession No. ML040090281.

17.   Yegorova, L., K. Lioutov, N. Jouravkova, A. Konobeev, V. Smirnov, V. Chesanov, and A. Goryachev, "Experimental Study of Embrittlement of Zr-1%Nb VVER Cladding under LOCA Relevant Conditions," NUREG/IA-0211, March 2005, ADAMS Accession No. ML051100343.

18.   Novikov, V.V., V.A. Markelov, V.N. Shishov, A.V. Tselishchev, and A.A. Balashov, "Results of Post-Irradiation Examinations (PIE) of E110 Claddings and Alloy Upgrading for VVER," *Transactions of the 2006 International Meeting on Light-Water Reactor Fuel Performance (Top-Fuel)*, European Nuclear Society, Salamanca, Spain, pp. 590–594, October 2006.

19.   Billone, M.C., and R.O. Meyer, "Comments on Proposed Rulemaking on 10 CFR 50.46 (ECCS Acceptance Criteria)," letter to Secretary, U.S. Nuclear Regulatory Commission, October 6, 2009, ADAMS Accession No. ML092800345.

20.   Oberländer, B.C., M. Espeland, and H.K. Jenssen, "LOCA Testing of High Burnup PWR Fuel in the HBWR, Additional PIE on the Cladding of the Segment 650-5," Institute for Energy Technology, IFE/KR/E-2008/004 (dated April 2008), June 2008, ADAMS Accession No. ML081750715.

21.   Karb, E.H., M. Prüssmann, L. Sepold, P. Hofmann, and G. Schanz, "LWR Fuel Rod Behavior in the FR2 In-pile Tests Simulating the Heatup Phase of a LOCA," Karlsruhe Nuclear Research Center, KfK-3346, March 1983.

22.   Oberländer, B.C., et al., "LOCA IFA650-4: Fuel Relocation Study," Institute for Energy Technology, Halden Reactor Project, Enlarged Halden Program Group Meeting, March 19, 2007, ADAMS Accession No. ML101600649.

NRC FORM 335
(12-2010)
NRCMD 3.7

U.S. NUCLEAR REGULATORY COMMISSION

BIBLIOGRAPHIC DATA SHEET

(See instructions on the reverse)

1. REPORT NUMBER
(Assigned by NRC, Add Vol., Supp., Rev., and Addendum Numbers, if any.)

NUREG/KM-0004

2. TITLE AND SUBTITLE

Fuel Behavior under Abnormal Conditions

3. DATE REPORT PUBLISHED

MONTH	YEAR
January	2013

4. FIN OR GRANT NUMBER

5. AUTHOR(S)

R.O. Meyer

6. TYPE OF REPORT

Technical

7. PERIOD COVERED (Inclusive Dates)

8. PERFORMING ORGANIZATION - NAME AND ADDRESS (If NRC, provide Division, Office or Region, U. S. Nuclear Regulatory Commission, and mailing address; if contractor, provide name and mailing address.)

Division of Systems Analysis
Office of Nuclear Regulatory Research
U.S. Nuclear Regulatory Commission
Washington, DC 20555-0001

9. SPONSORING ORGANIZATION - NAME AND ADDRESS (If NRC, type "Same as above", if contractor, provide NRC Division, Office or Region, U. S. Nuclear Regulatory Commission, and mailing address.)

Same as above

10. SUPPLEMENTARY NOTES

11. ABSTRACT (200 words or less)

Under normal operating conditions, cladding and core structural materials operate around 300 degrees Celsius (C), and fuel pellets experience peak temperatures below 2,000 degrees C at the pellet centerline. At these temperatures, fuel cladding integrity is maintained, and fission products are contained within the fuel rods. However, under some abnormal conditions, higher temperatures and other conditions significantly alter the behavior of these materials. These conditions can threaten core coolability and lead to fission product release. This NUREG report considers the following two types of accident conditions: (1) reactivity-initiated accidents and (2) loss-of-coolant accidents. This report describes the fuel behavior of each accident condition from basic concepts to the current state of the art. It also mentions safety criteria, references the classic experimental work in each of these areas, and presents equations and figures that permit some quantitative evaluations.

12. KEY WORDS/DESCRIPTORS (List words or phrases that will assist researchers in locating the report.)

Fuel, Cladding, Zirconium, Hydrogen , Loss-of-Coolant Accident (LOCA), Reactivity-Initiated Accident (RIA)

13. AVAILABILITY STATEMENT

unlimited

14. SECURITY CLASSIFICATION

(This Page)

unclassified

(This Report)

unclassified

15. NUMBER OF PAGES

16. PRICE

NRC FORM 335 (12-2010)

NUREG/KM-0004
January 2013